Emerging Free and Open Source Software Practices

Sulayman K. Sowe
Aristotle University of Thessaloniki, Greece

Ioannis G. Stamelos
Aristotle University of Thessaloniki, Greece

Ioannis M. Samoladas
Aristotle University of Thessaloniki, Greece

IGI PUBLISHING

Hershey • New York

Acquisition Editor:	Kristin Klinger
Senior Managing Editor:	Jennifer Neidig
Managing Editor:	Sara Reed
Assistant Managing Editor:	Sharon Berger
Development Editor:	Kristin Roth
Copy Editor:	Mike Goldberg
Typesetter:	Diane Huskinson
Cover Design:	Lisa Tosheff
Printed at:	Yurchak Printing Inc.

Published in the United States of America by
IGI Publishing (an imprint of IGI Global)
701 E. Chocolate Avenue
Hershey PA 17033
Tel: 717-533-8845
Fax: 717-533-8661
E-mail: cust@igi-pub.com
Web site: http://www.igi-pub.com

and in the United Kingdom by
IGI Publishing (an imprint of IGI Global)
3 Henrietta Street
Covent Garden
London WC2E 8LU
Tel: 44 20 7240 0856
Fax: 44 20 7379 0609
Web site: http://www.eurospanonline.com

Library of Congress Cataloging-in-Publication Data

Emerging free and open source software practices / Sulayman K. Sowe, Ioannis G. Stamelos and Ioannis Samoladas, editors.
  p. cm.
  Summary: "This book is a collection of empirical research regarding the status of F/OSS projects, presenting a framework and state-of-the-art references on F/OSS projects, reporting on case studies covering a wide range of F/OSS applications and domains. It asserts trends in the evolution of software practices and solutions to the challenges ubiquitous nature free and open source software provides"--Provided by publisher.
  Includes bibliographical references and index.
  ISBN 978-1-59904-210-7 (hardcover) -- ISBN 978-1-59904-212-1 (ebook)
  1. Open source software. 2. Computer software--Development. I. Sowe, Sulayman K. II. Stamelos, Ioannis, 1959- III. Samoladas, Ioannis.
  QA76.76.S46E44 2007
  005.3--dc22
                              2007007268

British Cataloguing in Publication Data
A Cataloguing in Publication record for this book is available from the British Library.

Emerging Free and Open Source Software Practices

Table of Contents

Section I:
Empirical Research into F/OSS

> *Martin Michlmayr, University of Cambridge, UK*
> *Gregorio Robles, Universidad Rey Juan Carlos, Spain*
> *Jesus M. Gonzalez-Barahona, Universidad Rey Juan Carlos, Spain*

> *Stefan Koch, Vienna University of Economics and Business Administration,*
> *Austria*

Foreword

The IBM 701, announced to the public in 1952, was IBM's first commercial scientific computer. Three years later, representatives from seventeen of its installations met to see how they could coordinate their programming work. The organization they formed was named SHARE, and this book contains parts of its DNA. Not many know that a key contribution of SHARE was the distributed development of system and application programs and their dissemination in source code form.

We've come a long way since the early days of SHARE. At that time, both IBM's operating systems and SHARE's patches and libraries were usually distributed in source code format. Nowadays, software distribution is bifurcated into two large branches. Software sold for profit is more often than not delivered in a closed binary format, while software available in open source format is typically distributed for free. This second branch is where we've seen significant growth and innovation in the last years. Auspiciously, this book examines the profound implications of the open source software development practices.

In a *Communications of the ACM* article titled "Software is not a Product," Phillip Armour fittingly described software as a medium for storing knowledge. Knowledge storage isn't novel: the DNA, our brain, material artifacts, and books have served this purpose for centuries or millennia. Yet software is different, because it is the first human-created mechanism that can actively execute this stored knowledge. And open source software is different again, because it allows everybody to examine both how the knowledge is derived (the process) and how it is expressed (the product).

This book's authors deal with many interesting aspects of the open source development process and the corresponding products. On the process front they examine the software development tools and human resource management methods that allow open source products to flourish against significant odds. As the records of the development process are typically also open, other chapters use those as a study field looking for evidence of quality, examining the dynamics of the project community, or using them as a test bed for performing process optimization experiments.

Further chapters focus on the open source products. Economies of scale often force proprietary software development companies to focus on the fat head of their potential users' power law distribution. Yet the Internet delivery and collaboration channels coupled with a development model based on volunteers allows open source products to occupy spectacular niches in the users' distribution—the famous "long tail." There are chapters examining life in that long tail and the ecosystem in which open source projects prosper.

Finally, there are chapters navigating the turbulent waters where open source, proprietary software, and the real world meet. How can we best adopt open source software in corporate environments? How can we handle migration from proprietary solutions? How are business models based on open source software related with the software's quality? Answering these questions isn't easy, but venturing on these directions is certainly fascinating.

SHARE's founders chose their organization's name hoping they could subsequently come up with a clever set of words to match the initials. None of the suggestions put forward were good enough, and therefore, true to today's open source spirit, they decided that individual members were free to interpret them in their own way. The 1956 *SHARE Reference Manual,* perhaps the first codification of open source principles and practices, notes that this action was symbolic of one of SHARE's core values: unity in essentials and freedom in the accidentals.

Diomidis Spinellis

Athens University of Economics and Business

Department of Management Science and Technology

http://www.dmst.aueb.gr/dds

Diomidis Spinellis is an associate professor at the Department of Management Science and Technology at the Athens University of Economics and Business, Greece. His research interests include software engineering tools, programming languages, and computer security. He holds an MEng in software engineering and a PhD in computer science both from Imperial College London. He has published more than 100 technical papers in the areas of software engineering, information security, and ubiquitous computing. He has also written the two "Open Source Perspective" books: Code Reading *(Software Development Productivity Award 2004), and* Code Quality. *He is a member of the IEEE Software editorial board, authoring the regular "Tools of the Trade" column. Dr. Spinellis is a FreeBSD committer and the author of a number of open source software packages, libraries, and tools.*

Dr. Spinellis is a member of the ACM, the IEEE, the Usenix Association, the Greek Computer Society, the Technical Chamber of Greece, and a founding member of the Greek Internet User's Society. He is a co-recipient of the Usenix Association 1993 Lifetime Achievement Award.

Preface

There is little doubt about the profound effect the Internet has on everyday life, but the symbiotic relationship between the Internet and free and open source software (F/OSS) will be the subject of interest for many years to come. The emergence of F/OSS continues to have great impact in the way we develop, support, maintain, and distribute software. F/OSS is collaboratively built software that is shared by developers and users and can be "freely" downloaded with or without the source code for use, modification, and further distribution. Prolific licensing agreements such as the General Public License (GPL) define the rights users have over the product. F/OSS participants develop software online, relying on extensive peer collaboration through the Internet, and are motivated by a combination of intrinsic and extrinsic motives. The *Bazaar,* as opposed to the *Cathedral*, model has emerged as a viable software development approach and has produced a number of successful applications in the areas of operating systems (Linux), Web services (Apache), database applications (MySQL, PostgreSQL), and so forth. These successes may be attributed to the idiosyncratic practices adopted by F/OSS projects, which are different from the common practices in traditional software engineering projects. This *new* way of developing software continues to attract the research interest of many individuals and organizations.

Trends in research findings coupled with the popularity of F/OSS have highlighted a number of *challenges* the development model and projects face. These challenges range from issues of inactive volunteerism, quality assurance, changes in the ecology and dynamics of the communities, effective tools and practices to manage and support the software development process, the lack of explicit guidelines and specifications inherent in traditional software development, and knowledge management issues, to the technicalities of how to meet the needs of a new generation of F/OSS developers, testers, and users. F/OSS projects have adopted practices to address some of these critical issues and help them succeed. On the other hand, a large number of projects never made it beyond the first version. However, each project is distinct and each hosting platform or portal poses unique challenges projects must surmount to succeed.

Beyond the academic and geek or *hacker* communities, F/OSS is making a tremendous impact in the way people work in private and public administrations and is redefining the software industry. The increased adoption of F/OSS by well-known organizations has led to more enterprise recognition and hence consideration. What is more compelling is that companies are increasingly implementing *open source strategies*—porting programs and applications into the Linux environment while at the same time realizing that they can charge complementary services such as post-sale services. The LAMP (Linux/Apache/MySQL/PHP or Perl) stack is one example of cost savings F/OSS offers. However, a growing number of companies take a cautionary approach towards F/OSS full adoption; the main challenges being reliability, total cost of ownership, ability to integrate with existing systems, performance, and scalability.

Targets and Overall Objective of the Book

Our understanding of some of the socio-technical, economical, and managerial issues surrounding F/OSS continues to increase but information on these issues remains sparse. This book leverages the expertise of authors from different parts of the world and is a compilation of software practices adopted in various F/OSS environments. Rather than being the work of one author, the book is a collection of edited chapters written by experts and practitioners in the field of software engineering, economics, sociology, mathematics, and so forth. The editors received twenty seven chapters initially, but after an extensive peer-review process, only twelve chapters were selected for inclusion in this book. The diversity in thoughts and experiences of the authors represents a joint work and cooperative action approach to F/OSS. The subject area is wide and entails many facets that the book attempts to cover:

- Empirical research on emerging F/OSS practices
- Discussion about the coordination and collaboration strategies used by various F/OSS communities
- Presentations on the tools both software users and developers use to produce quality software
- Adoption of F/OSS in public and corporate environments
- Case studies of successful and failed F/OSS projects

The book chapters combine techniques with practice to establish a synergistic view of F/OSS so that it will benefit academicians, researchers, managers, policy makers, and F/OSS developers and users at large. However, unlike its predecessors, the book does NOT begin with a concise history of F/OSS or address the ideological issue of terminology. Different terms appear in the chapters:

- Open source software (OSS)
- Free open source software (F/OSS)
- Free/libre and open source software (F/LOSS)
- Libre software

In the spirit of the domain, what F/OSS is and is not has been left to the individual authors to give their own opinion.

Sulayman K. Sowe
Co-Editor
Department of Informatics
Aristotle University, Greece
2007

Organization of the Book

The book is organized into twelve chapters. A brief description of each of the chapters follows:

Chapter I presents an analysis of the evolution over time of the human resources in the Debian project and discusses, in detail, how Debian handles the volatility of the volunteers who made it happen. The authors of this chapter contend that F/OSS volunteer contributions are inherently difficult to predict, plan, and manage and yet F/OSS projects, large and small, rely on volunteers. The fairly unstructured collaboration of volunteers has demonstrated itself to be a viable software development strategy, even if it is associated with certain challenges related to project management and software quality. The reliance on volunteer contributions leads to some fundamental management challenges in most large-scale F/OSS projects.

Chapter II proposes for the first time a method to compare the efficiency of F/OSS projects, based on the data envelopment analysis (DEA) methodology. The author argues that DEA offers several advantages in this context, as it is a non-parametric optimization method without the need for the user to define any relations between different factors or a production function. He further suggests that the methodology can account for economies or diseconomies of scale, and is able to deal with multi-input, multi-output systems in which the factors have different scales. Using a data set of 43 large F/OS projects retrieved from SourceForge.net, the author demonstrates the application of DEA. Analysis of the results shows that DEA indeed is usable for comparing the efficiency of projects in transforming inputs into outputs, that differences in efficiency exist, and that single projects can be judged and ranked accordingly.

Chapter III documents the evolution and ecology of communities within the Apache project. The authors describe community structures in terms of social networks measures such as community size or degree distribution. They further explore potential growth factors and mechanisms that help explain the observed evolution of the communities. Their hypotheses about the evolution of a portfolio (or ecology) of the communities are validated through an empirical analysis of eight years of the Apache project's mailing list archives.

Chapter IV presents a holistic view, and investigates the social aspects, of F/OSS communities by first analyzing existing literature on the motivation of F/OSS community participants, and then presenting the results of survey research on community participants' perceptions of their own participation. By collecting and analyzing qualitative and quantitative data, the authors demonstrate that F/OSS communities are not just confined to support and development but are hosts to many other activities, including the promotion of F/OSS, business and personal development, and educational activities. Furthermore, it is not just source code that is openly shared but also knowledge, skills, ideas, and expertise. The authors conclude that it is the transference and dissemination of these, between a diverse set of participants, which allows F/OSS communities to function so effectively.

Chapter V utilizes exploratory data analysis of F/OSS projects' documents, Web sites, mailing lists, and a review of current literature to study F/OSS communities. The chapter sets the scene for the ways in which developers and users coordinate their activities to create software. The authors discuss what F/OSS communities are and how they operate, and how organizations adopting F/OSS interact with the developers in the communities developing the software. Coordination, the authors contend, is more than getting individuals motivated to do the work they do; it also means arranging and ordering individuals' efforts, it refers to labour division and to task specialization. Furthermore, face-to-face, semi-structured interviews with experts in The Netherlands, the U.S., and Germany are presented to better understand coordination in F/OSS communities. The authors enumerate a number of mechanisms and tools that support the processes of collaboration.

Chapter VI provides a starting point for discussion between the F/OSS community and end users about software quality. One important issue highlighted by the chapter is that the potential benefits associated with diversity in F/OSS come at a price, as the donated code can be of varying quality. The authors suggest that F/OSS communities need to develop appropriate mechanisms to signal quality to end users in simple, easy to understand terms. By providing metrics for validating the quality of the F/OSS artifacts, the chapter models the overall F/OSS development process using soft systems methodology and UML. The chapter also describes how such metrics could be used to improve the development process, and demonstrate how automated extraction of such metrics could be employed alongside F/OSS repositories such as Sourceforge.net. An illustrative example is used to demonstrate the practical applicability of the proposed approach, with a description of an early prototype of the supporting information system.

Chapter VII presents a case that the future success of F/OSS is not based on the fact that there is no associated licensing fee, but on the question of the quality of the software. The authors explore the importance software quality has for software developers and discusses the conditions needed for software quality to be realized. Doing this, the authors delve into both the motivations of individual programmers and the business models of firms that sponsor F/OSS projects. While the motivations of programmers might be different from the motivations of the firms employing them, the authors argue that companies adopting F/OSS can benefit from such diverse opinions. If we understand the people and their motivations for developing F/OSS, the authors conclude that this might increase our confidence in the quality and sustainability of F/OSS and, as a result, lower the barrier to use it.

Chapter VIII seeks to identify and characterize the array of social and technical resources needed to support the development of F/OSS supporting e-commerce (EC) or e-business (EB) capabilities. The author reports on a case study within a virtual organization that has undertaken an organizational initiative to develop, deploy, and support F/OSS systems for EC or EB services. The case study of the GNUe arises from a longitudinal field study spanning four years and employing grounded theory techniques including axial coding and construction of comparative memoranda based on field data collected through face-to-face and e-mail interviews, as well as extensive collection and cross-coding of publicly available project documents and software development artifacts posted on the project's Web site. The chapter identifies many types of socio-technical resources and resource-based capabilities for Free EC/EB that may explain/predict (a) what's involved, (b) how it works, or (c) what conditions may shape the longer-term success or failure of such efforts. The study links F/OSS with enterprise resource planning (ERP) and EC/EB.

Chapter IX made a number of recommendations and present lessons learned from the authors' research conducted on the migration of PAs to open source desktop software (OpenOffice.org). They report on the migration of the Government of the Brussels-Capital Region towards OpenOffice.org and integrate and compare their work to various studies on the migration of PAs towards F/OSS on the desktop. The authors present a set of best practices—based on empirical research—for the migration towards a F/OSS desktop environment.

Chapter X draws on the authors' experience to present the technical and managerial issues SMEs and PAs should consider when introducing a GNU/Linux-based desktop system into their corporate environment. The authors contend that for many small and medium enterprises (SMEs) and public administrations (PAs), the introduction of a GNU/Linux-based desktop system is often problematic. The technical obstacles are represented by different hardware configurations that might require several ad-hoc activities to adapt a standard GNU/Linux distribution to the specific environment. Managerial issues are related to employees' training costs. To address these issues, the authors present DSS (Debased Scripts Set)—a next-generation live distribution that includes an unmodified Debian-based Linux release and a modular-designed file system with some extended features.

Chapter XI posits that customers who deploy technology products and/or platforms that fail to attract a thriving ecosystem or whose ecosystem deteriorates are increasingly faced with declining availability of skills, increasing operating costs, and/or lower levels of innovation. The author presents a framework and case study to highlight a number of characteristics and aspects of F/OSS projects as ecosystems. The author further characterizes individuals and organizations as members of a larger ecosystem that includes the original vendor as well as supporting foundations, external service partners, integrators, distributors, and the users themselves. The framework developed in this chapter is used to examine how such ecosystems can be evaluated by existing and potential members to gauge the health and sustainability of projects and the products and services they produce. The author's case study is based on the open source-based Evergreen project which is sponsored by the Georgia Public Library Service (GPLS).

Chapter XII concludes the book and presents the author's first-hand experience report about his involvement in the initiation, development, community activity coordination, and the final demise of an F/OSS project—Kalbum, which briefly flourished between October 2002 and March 2003. The story goes that, for every amazingly successful F/OSS project, there is a Kalbum somewhere, lurking in the shadows. The experience report is that of a sole developer who had to fight for time with work and family commitments working in a project which did not inherit its codebase from one of the largest, and incumbent community-developed projects in existence. If there is anything common to Kalbum and other projects, it is the F/OSS development model. The way the case study is presented is rather unusual for academic research, but the author documents the story of many untold and "unsuccessful" projects hosted at various F/OSS portals (e.g., sourceforge.net).

Acknowledgment

The editors would like to sincerely acknowledge the unflinching support of all individuals involved in the collation and review process of the book's chapters, without whose support the project could not have been satisfactorily completed. However, some of the authors of chapters not included in this book also served as referees for included chapters. Thanks go to all those reviewers for their constructive and comprehensive reviews.

Special thanks also go to all the staff at IGI Global, whose contributions throughout the whole process from inception of the initial idea to final publication have been invaluable. In particular, we wish to extend our sincere gratitude to *Kristin Roth*, with whom we started this project, and *Ross Miller*, who collaborated with us to final publication. Their cooperation in the support and development aspects of this book has been outstanding.

In closing, we want to thank the research teams of the PLaSE lab (http://plase. csd.auth.gr/) and the Software Engineering Group (http://sweng.csd.auth.gr/) for the provision of facilities and moral support which make this work possible, the Department of State of Education of The Gambia and the Greek State Scholarships Foundation (IKY) for their support. This work was partially funded by the European Community's Sixth Framework Programme under the contract IST-2005-033331 "Software Quality Observatory for Open Source Software" (SQO-OSS) (http://www. sqo-oss.eu). Our final thanks go to the software developers and users, researchers, and all proponents in the free/open source community for making this emerging software development methodology happen and ubiquitous in the future.

Sulayman K. Sowe
Co-Editor
Department of Informatics
Aristotle University, Greece
2007

Section I

Empirical Research into F/OSS

Chapter I

Volunteers in Large Libre Software Projects:
A Quantitative Analysis Over Time

Martin Michlmayr, University of Cambridge, UK

Gregorio Robles, Universidad Rey Juan Carlos, Spain

Jesus M. Gonzalez-Barahona, Universidad Rey Juan Carlos, Spain

Abstract

Most libre (free, open source) software projects rely on the work of volunteers. Therefore, attracting people who contribute their time and technical skills is of paramount importance, both in technical and economic terms. This reliance on volunteers leads to some fundamental management challenges: Volunteer contributions are inherently difficult to predict, plan, and manage, especially in the case of large projects. In this chapter we present an analysis of the evolution over time of the human resources in large libre software projects, using the Debian project, one of the largest and most complex libre software projects based mainly in voluntary work, as a case study. We have performed a quantitative investigation of data corresponding to roughly seven years, studying how volunteer involvement has affected the software released by the project, and the developer community itself.

Introduction

Volunteer contributions are the basis of most libre[1] software projects. However, the characteristics, and the way of working of volunteers, can be quite different from those of employees who are the main force behind traditional software development. Volunteers can contribute with the amount of effort they want, can commit for the time period they consider convenient, and can devote their time to the tasks they may prefer, if the context of the project makes that possible (Michlmayr & Hill, 2003). But even in this apparently difficult environment, many libre software projects have produced systems with enough quality and functionality to gain significant popularity. Therefore, the fairly unstructured collaboration of volunteers has been demonstrated as a viable software development strategy, even if it is associated with certain challenges related to project management and quality (Michlmayr, 2004). In this chapter we explore how these voluntary contributions evolve over time in one of the largest libre software projects, Debian.

For our purposes, we will define volunteers as those who collaborate in libre software projects in their spare time, not profiting economically in a direct way from their effort. Volunteers can be professionals related to information technologies, but in that case their activity in the libre software project is not done as a part of their professional activity. Although the vast majority of participants in libre software projects comply with our definition, it is important to note that there are also non-volunteers, that is, paid people (normally hired or contracted), who produce libre software. German has studied paid employees from various companies in the GNOME project (German, 2004). He notes that they are usually responsible for less attractive tasks, such as project design and coordination, testing, documentation, and bug fixing. Also, "[m]ost of the paid developers in GNOME were, at some point, volunteers. Essentially for the volunteers, their hobby became their job."

The involvement of volunteers, of course, raises new economic and business model issues that have to be taken into account in commercial strategies around libre software. Collaboration from volunteers is difficult to predict, but if it is given, it may add value to a software system in very economic terms for a software company.

The structure of this chapter will be as follows. In the second section we discuss the nature and, in particular, the tasks performed by volunteers, paying special attention to those who contribute to Debian, the case study investigated in this chapter. Following this section, a set of research questions regarding volunteer participation will be raised. The primary goal of this chapter is to answer these questions based on quantitative data. The methodology for retrieving the quantitative data used in this study is first given. In this section, we also propose a number of measures that will allow us to answer the questions. The results we have obtained as part of this study will be presented and commented on in depth for the Debian project. Finally, conclusions, applicability of the methodology, and further research will be discussed.

The Debian Project and its Volunteers

Debian is an operating system completely based on libre software (Monga, 2004; O'Mahony, 2003). It includes a large number of applications, such as the GNU tools and Mozilla, and the system is known for its solid integration of different software components. Debian's most popular distribution, Debian GNU/Linux, is based on the Linux kernel. Ports to other kernels, such as Hurd and FreeBSD, are in development.

One of the main characteristics of the Debian distribution is that during the whole life of the project it has been maintained by a group of volunteers, which has grown to a substantial number. These individuals devote their own time and technical skills to the creation and integration of software packages, trying to supply users with a robust system which provides a lot of functionality and technical features.

Following our definition of volunteers, all maintainers in Debian are volunteers. Some employers of people who act as Debian maintainers in their spare time permit their staff to devote some of their time to Debian during work hours. Nevertheless, the majority of work by most Debian maintainers is performed in their spare time. In contrast to some projects, such as the Linux kernel and GNOME, there are no Debian maintainers who are paid to work on the system fulltime, even though a number of organizations have a commercial interest in Debian and contribute varying degrees of manpower (and other resources) to the project. For example, a number of regions in Spain have their own operating systems based on Debian; HP made Debian more suitable for large telecom customers, and Credativ provides commercial services for Debian and similar systems.

There are several tasks that volunteers can do in Debian: maintaining software packages, supporting the server infrastructure, developing Debian-specific software, for instance, the installation routine and package management tool, translating documentation and Web pages, and so forth. From all these tasks, we will focus in this chapter on package maintainers, whose task it is to take existing libre software packages and to create a ready-to-install Debian package. Debian maintainers are also called Debian developers, although their task is really not to develop software but to take already developed software for the creation of a Debian package. This, of course, does not mean that a Debian maintainer may not develop and maintain software, but this is not usually the case: the original author (or developer), known as the 'upstream' developer, and the Debian maintainer, are usually, but not necessarily, not the same person.

Besides its voluntary nature, the Debian project is unique among libre software projects because of its social contract (Debian Social Contract, 2006). This document contains not only the primary goals of the Debian project, but also makes several promises to its users. Additionally, there are a number of documents Debian maintainers have to follow in order to assure quality, stability, and security of the resulting distribution. In particular, Debian's Policy document ensures that the large number of volunteers working independently will produce a well-integrated system rather than merely an aggregation of software packages which do not play together very well (Garzarelli & Galoppini, 2003).

There has recently been some interest in studying how the voluntary status of Debian members affects the quality of the resulting product. Managing volunteer contributors is associated with certain problems that "traditional" software development usually does not confront (Michlmayr & Hill, 2003). It is known that there are some intrinsic problems when the development process is carried out in a distributed fashion (Herbsleb et al., 2001). The situation in Debian and similar projects is even more complex because the development process is not only distributed but also largely based on volunteers. This can lead to certain challenges, such as the unpredictability of the level of their involvement (Michlmayr, 2004).

To some degree, the volatility of voluntary contributors can be limited by the introduction of more redundancy, such as the creation of maintainer teams. The creation of teams and committees for specific purposes, such as management, or for complex tasks has been already reported in other libre software projects (as for instance, German's work on the GNOME project [2004]).

Research Objectives and Goals

Related research has been very active in studying the static picture of a libre software project community over the last years, as can be seen from studies performed on Apache and Mozilla (Mockus et al., 2002), FreeBSD (Dinh-Trong & Bieman, 2005) or GNOME (Koch & Schneider, 2002), leading to models that discuss onion-like structures of libre software projects (Crowston & Howison, 2005). The main goal of this chapter is to introduce the time axis in these kinds of studies, focusing most notably on the contribution of volunteers.

We have therefore set up a list of research questions which we would like to answer for several large libre software projects. In the case of Debian, we have additional information that permits us to link the work done by a volunteer with a piece of software, so we can, for example, study what has happened to packages from volunteers who have left the project. In the following, the set of questions that we are raising will be answered in detail for the Debian project.

The specific questions we aim to answer with this chapter are the following:

a. **How many volunteers does the project have, and how does this number change over time?** This will provide us with some basic data, useful when working with subsequent questions. When we started the study, we expected a steady increase of volunteers over time, as it is already known that the number of packages included in the system has been growing in that way (Gonzalez-Barahona et al., 2004). In addition, we will try to find out if the work ratio (measured as activity or output per developer) has increased over time or not.

b. **How many volunteers from previous releases remain active?** We want to measure the volatility of the volunteers in large libre software projects. That is, do volunteers join the project and work on it for short periods of time, or, on the contrary, do they stay for many years? Specifically, we want to calculate the half-life of contributors of the project. The half-life is defined as the time required for a certain population of maintainers to fall to half of its initial size. This figure could be easily compared with other libre software projects and, of course, with statistics from companies from software and other industries.

c. **What is the activity of volunteers who remain in subsequent releases?** Answering this question will allow us to know if "older" volunteers strengthen their contributions as time passes, contributing more to the project, or whether they become less active. There are two possible hypotheses one could propose. On the one hand, those volunteers who have been involved for a long time may be very experienced and therefore more efficient in their work than less experienced developers. On the other hand, young developers may have more time or energy to devote to the project and therefore contribute more. Both theories are possible and mutually compatible.

d. **What happens to packages maintained by volunteers who leave the project?** Our intention is to see if we can find a regeneration process in libre software projects that allows them to survive the loss of some of their human resources.

Since Debian maintainers are volunteers, they may quit the project at almost any time, leaving their packages unmaintained. There are two possible outcomes regarding the future of those packages: First, they can be taken over (adopted) by another maintainer. Alternatively, if nobody is interested in adopting them, they will eventually be removed from the archive and excluded from future stable releases. Such removals of unmaintained packages are part of Debian's Quality Assurance effort. Our intention was to know how this inherent characteristic of the voluntary contributors affects Debian, and how this is damped down by other (possibly new) maintainers.

e. **Are more "important" and commonly used packages maintained by more experienced maintainers?** It can be interesting to know whether packages which are considered crucial for the functioning of the system are maintained primarily by volunteers who have more experience. For this, we have considered the most used packages as the targets of the study. We have defined as crucial packages those which are usually installed on every system, as, for instance, the base system which in the case of the Debian GNU/Linux operating system is composed, among others, of the Linux kernel and the GNU tools. This does not necessarily mean, of course, that crucial packages are more difficult to maintain than other packages, but as they are used by all users of the system and the rest of the software heavily depends on their proper functioning, these packages have to be maintained with special care. Data about the importance of each package will be obtained from the Debian Popularity Contest[2] that tracks how many people have installed a given package.

Methodology and Sources of Data

Debian consists of four parallel versions (stable, testing, unstable, experimental) which can be downloaded from the Internet. The focus of this study is on the stable versions from Debian 2.0 onwards, up to version 3.1, which provide good snapshots of the history of the distribution. These releases comprise a period of time from July 1998 to June 2005. There have been releases of Debian before 1998 (Lameter, 2002), but they have not been taken into consideration for this study since the sources of data we have used in this study were not available for them. For each release, we have retrieved the corresponding Sources.gz file (see next section) from the Debian archive. We have then extracted information about packages and their maintainers from this file and stored the results in a database. After that, we performed some semi-automatic cleaning and massaging of the data that will be explained in more detail below. Final results were obtained through queries to the database, and correlations that have been implemented by another set of scripts.[3]

In addition to the analysis of official releases, we have enriched the findings by additionally taking a more fine-grained data source into account. While releases are only done occasionally, in the case of Debian with years between releases, uploads to the Debian archive are done on a continuous basis. We have analyzed the activity related to these uploads to clarify some of the findings of the paper on which this chapter is based (Robles et al., 2005). The estimations of the size of the releases have been done using a software that counts source lines of code and avoids double-counting the code included in various packages (this methodology is described in detail (Gonzalez-Barahona et al., 2001)), using previously published data (Gonzalez-Barahona et al., 2004), except for release 3.1, which was calculated specifically for this study. The data related to the importance of packages has been retrieved from the Debian Popularity Contest (see section on this topic).

Debian Sources File

Since version 2.0, the Debian repository contains a Sources.gz file for each release, listing information about every source package in it. Every source package contains the name and version, list of binary packages built from it, name and e-mail address of the maintainer, and some other information which is not relevant for this study. As an example, see an excerpt of the entry for the mozilla source package in Debian 2.2 below. It can be seen how it corresponds to version M18-3, provides four binary packages, and is maintained by Frank Belew.

```
[...]
Package: mozilla
Binary: mozilla, mozilla-dev, libnspr4, libnspr4-dev
Version: M18-3
Priority: optional
Section: web
Maintainer: Frank Belew (Myth) <frb@debian.org>
Architecture: any
Standards-Version: 3.2.0
Format: 1.0
Directory: dists/potato/main/source/web
Files: 57ee230b97ccc69444ccccd0bc66908a 719 mozilla_M18-3.dsc
532934635ad426255036ee070bad03c8 28642415 mozilla_M18.orig.tar.gz
3adf83de7e74bf940ee02c0deca20372 18277 mozilla_M18-3.diff.gz
[...]
```

Debian Popularity Contest

The Debian Popularity Contest is an attempt to map the usage of Debian packages. Its main goal is to know what software packages are actually installed and used. Information from the Popularity Contest is used by Debian, for example, to decide which software to put on the first CD.

The system functions as follows: Debian users may install the popcon package, which sends a message every week with the list of packages installed on the machine as well as the access time of some files which may give hints regarding the last usage of these packages. Of course, privacy issues are considered in a number of ways: Upon installation, users are explicitly asked if they want to send this information to Debian, and the server which collects the data anonymizes it as much as possible.

Table 1. Debian Popularity Contest statistics

rank	name	inst	vote	old	recent	no-files	(maintainer)
1	adduser	6881	6471	94	316	0	(Adduser Development)
2	debianutils	6881	6517	50	314	0	(Clint Adams)
3	diff	6881	6425	261	195	0	(Santiago Vila)
4	e2fsprogs	6881	5448	825	608	0	(Theodore Y. Ts'o)
5	findutils	6881	6449	233	199	0	(Andreas Metzler)
6	grep	6881	6436	126	319	0	(Ryan M. Golbeck)
7	gzip	6881	6558	245	78	0	(Bdale Garbee)
8	hostname	6881	6112	715	54	0	(Graham Wilson)
9	login	6881	6407	56	418	0	(Karl Ramm)
10	ncurses-base	6881	56	143	6	6676	(Daniel Jacobowitz)

The resulting statistical information of all users participating in this scheme is publicly available on the Web site of the project. For every package, it includes the number of machines on which it is installed (inst), the number of machines which make regular use of that package (vote), the number of machines with old versions of the package (old), the number of recent updates (recent), the number of machines where not enough information is available (no-file), and the maintainer of the package. Below is an excerpt of the available data, in this case the top ten packages ordered by installations as of December 4, 2004. The first 66 packages are installed on all machines, with 6881 installations.

Debian Developer Database

From June 1999 onwards, Debian has held a database (http://db.debian.org) with data related to members of the project. Some information, such as the full name and user name, can be retrieved publicly through the Internet. This database also contains information about the digital keys used by a developer. Debian makes use of digital signatures through the use of the tools PGP (Pretty Good Privacy) and GPG (GNU Privacy Guard) to approve uploads to their software archive. The use of digital signatures provides two guarantees. First, the signature will show that the package comes from a trusted source, that is, from an official Debian developer whose PGP or GPG key is stored in the Debian keyring and this developer database. Second, by verifying the signature on the package, it can be ensured that the package has not been tampered with during the process of uploading it to the Debian archive.

Package Uploads

While the main data source of this chapter is the Sources.gz files from the last official Debian releases, we have additionally taken uploads to the archive into account to answer some of the research questions with more detail. As mentioned above, Debian's archive is separated into various branches. The official releases are known as the stable branch, whereas Debian's development tree is known as "unstable." Even though it is said that the development tree is usually fairly stable, it is where major development occurs and as such, major bugs are introduced from time to time.

When an upload is made to unstable, a summary of the changes is automatically sent to a mailing list known as "debian-devel-changes." By extracting data from the archives of this list, the uploads made to the Debian archive over the last few years can be studied. The archive of this mailing list starts towards the end of August 1997. Because August is not complete, we use the data starting with September, since having data for full months allows for better comparisons. We then divided the whole time period into periods of three months each, leading to 34 periods which cover 8.5 years (102 months), from September 1997 leading up to February 2006. In total, slightly more than 181,500 uploads have been observed in this period, leading to a mean number of uploads of around 1,800 each month.

The method used to extract this information is as follows: first, the archives of the "debian-devel-changes" mailing list are downloaded from Debian. These are then parsed, leading to one message for each upload. These messages are signed by the developer's PGP or GPG key as described above. The information from these digital signatures is then used to map each upload to a unique user id corresponding to the developer who made the upload. The mapping from PGP or GPG key to user name is obtained from Debian's developer database and missing entries for old developers are manually supplemented. After this information is obtained, information about the developer (user name) and the date of each upload is stored in a database together with the message id from the posting stored in the archive.

The extraction of this data leads to fairly precise results but there are some limiting factors. First of all, it is important to take into account that uploads are not a measure of effort. We use the data as an indication of activity of a developer but the information is not rich enough to give specific information about how much effort was involved with a specific upload. Second, for the last few years Debian had the concept of "sponsorship," whereby an official Debian developer would upload a package created by a prospective developer (who is not part of Debian yet and whose GPG key is therefore not recognized). In such cases, the main effort was done by the prospective developer but the signature shows the name of the official Debian developer. Since we are concerned with activity of developers and not with effort, this is not an obstacle but it has to be considered during the interpretation of the data.

Finally, while the archive software used by Debian now sends automatic notifications of new uploads to the "debian-devel-changes" list, this was done manually in the past. Therefore, data from the past can be slightly unreliable. For example, we observed a number of messages which were not signed by PGP or GPG. Out of about 185,000, we have only detected about three thousand uploads for which no information about the developer could be extracted automatically. We believe that these are not significant and that data from uploads greatly enriches the findings from studies using the Source.gz file from official Debian releases.

Evolution of the Number of Debian Maintainers

The information of the evolution of the number of Debian maintainers will provide us with some basic data useful when working with subsequent research questions. When we started the study, we expected a steady increase of maintainers over time, as it is already known that the number of packages included in the system has been growing linearly (Gonzalez-Barahona et al., 2004). In fact, we expected the packages-to-maintainer ratio to be nearly constant, since it seems reasonable to consider that volunteers devote similar amounts of effort over time, which would lead to a constant number of packages per maintainer.

Figure 1 shows on the left side the evolution of the number of Debian maintainers for the latest five stable releases. As we have expected, the number of Debian individual maintainers has been growing over time. Debian 2.0 (July 1998) was put together by 216 individual maintainers, while the number of maintainers for later releases are 859 for 3.0 (July 2002) and 1,314 for 3.1 (June 2005). This shows a growth of about 35 percent every year. The right side of the figure shows the cumulative number of developers who have made uploads to Debian (starting as of September 1997 and finishing with February 2006). This chart gives more precise information as to the growth over time but it does not include some information which the other chart captures. As mentioned above, only official Debian developers are considered and therefore the numbers are lower than in the chart on the left side which considers all maintainers of packages in Debian, regardless of their official status. The more detailed information the chart on the right conveys is very interesting, though. A remarkable stagnation of the growth can be observed. This is because the New Maintainer process, Debian's admission process, was stopped for several months at the end of 1999. The growth continues in the middle of 2000 and, interestingly enough, the pause in the admission of the developers did not have any significant effect on the overall growth of the project.

Based on the data from official releases, we have conducted a small statistical analysis; the results are shown in Table 2. The ratio of packages per maintainer

Figure 1. Number of maintainers over time. The left chart is based on Debian re-leases while the right one is based on continuous uploads.

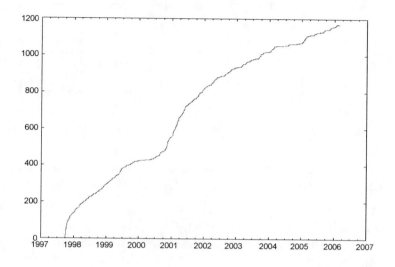

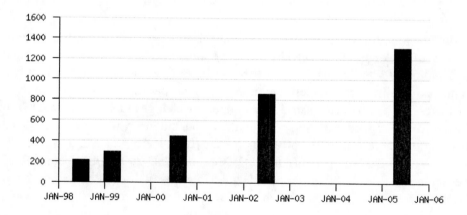

(see column "Pkg/Maint") grows over time, contrary to our initial hypothesis. The growth of packages is actually bigger than that of volunteers who contribute to the project. There are some possible explanations for this finding. First, it is possible that improvements of the development tools or in the practices employed have led

Table 2. Statistical analysis of the growth in number of Debian maintainers

Date	Release	Maint	Packages	Pkg/Maint	Median	Mode	Std. Dev	Gini	Max
Jul 98	2.0	216	1,101	5.1	3	1 (52)	5.8	0.492	50
Mar 99	2.1	296	1,559	5.3	3	1 (76)	6.5	0.521	55
Aug 00	2.2	453	2,601	5.7	3	1 (122)	7.4	0.535	69
Jul 02	3.0	859	5,119	6.0	4	1 (208)	8.2	0.539	79
Jun 05	3.1	1,314	7,989	6.1	3	1 (386)	9.1	0.577	127

to an increase in the efficiency of developers. Second, due to increased interest in libre software, the development speed in general has accelerated and volunteers are more committed.

Interesting enough, the median does not vary (with the exception of Debian 3.0) over time in these last years. Half of the maintainer population does not have more than three packages to maintain. Furthermore, the mode shows that the most frequent situation is a maintainer who is in charge of one package. In brackets we can find the number of developers who actually maintain only one package, which is around one forth of the total population of Debian maintainers.

The next three values (the standard deviation, the Gini coefficient,[4] and the maximum number of packages maintained by a single maintainer) strengthen the idea that the distribution of work tends to be distributed in a more unequal way, with a small number of maintainers maintaining more and more packages while the number of packages the vast majority is in charge of does not change much. Compared to other libre software applications and, in general, to other studies which have looked at the distribution of work in libre software projects (Ghosh &d Prakash, 2000; Koch & Schneider, 2002; Hunt & Johnson, 2002; Mockus et al., 2002; Ghosh et al., 2002), we can see that, unlike other projects, Debian is far away from a Pareto distribution. In terms of the Gini coefficient, Debian shows values from roughly 0.5 to 0.6 while studies of the activity in CVS repositories of other projects have found Gini to be in the range between 0.7 and 0.9 (Robles, 2006).

Finally, in Figure 2 we see the number of individual developers making uploads to the Debian archive for each three month period. A significant growth in the number of contributors can be seen in the first few years of the observed period. Since roughly the beginning of 2001, a fairly constant number of individuals makes contributions to Debian. There are about 550 unique contributors, even if they change over time.

The first column gives the date of the release specified in the second one. "Maint" is the number of maintainers that maintain at least a package, "Packages" the num-ber of total packages for that release, "Pkg/Maint," the mean number of packages per maintainer, "Median" the median number of packages, "Mode" gives the most

Figure 2. Mean number of developers making uploads per each three month period

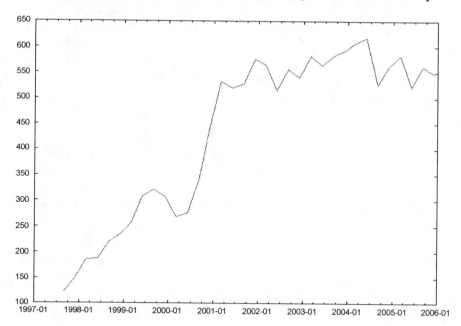

frequent contribution in number of packages and in brackets the number of maintainers who contribute to it, "Std. Dev," the standard deviation of our sample," "Gini" the Gini coefficient, and "Max" the maximum number of packages that a unique maintainer is responsible for.

Tracking Remaining Debian Maintainers

At the time of the release of Debian 2.0 in July 1998 there were 216 voluntary developers contributing to Debian. We have studied how the involvement of these 216 contributors to Debian 2.0 has changed over time. Table 3 gives an overview of the number of contributors from the original group left at each release, as well as the number of packages maintained by them. As the figure shows, the number decreases steadily, with only 117 of the original 216 contributors (54.2%) still possessing ownership of a package in June 2005. Based on these figures, we concluded in a previous paper that the half-life value had not been reached after six and a half years and estimated that the half-life value would be around 7.5 years (or 90 months) (Robles et al., 2005).

Taking the more fine-grained information from uploads into account, we can now revise these findings. Taking package ownership as an indication for activity is error prone, since it has been shown that a number of "maintainers" are actually inactive and do not maintain their packages (Michlmayr, 2004). It can take several months or longer until the situation is resolved, in particular if maintainers are busy but do not want to admit to themselves that they do not actually have enough time anymore. Uploads are therefore a much better measure in this case since they show activity. While this measure does not show the effort done by a maintainer it shows that they are still active, which is the question being asked here.

Based on data from uploads, we can see that as of the three-month period starting in June 1998, only 187 out of the original 216 contributors (which still possess packages in the release done in July 1998) are still active. Taking these 187 developers as the new population, we find that the group reaches their half-life in the periods between June and September 2004. This leads to a half-life value of less than 78 months (6.5 years). This is in line with the originally estimated value of 7.5 years, a slight over-prediction due to the fact that it takes some time until inactivity is reflected in the maintainer field of a package. The value obtained from this population is also in line with those obtained from other populations observed as part of the investigation of uploads to Debian, as can be seen in Figure 3. They all show a half-life of between 75 and 90 months.

It would be interesting to perform further analysis about which factors influence how long volunteers remain active. There is already evidence that some volunteers face feelings of burn-out (Hertel et al., 2003), but further studies into human-resource management and motivation in libre software projects could have positive effects on extending the half-life of volunteer contributions.

The number of packages for which these developers are responsible is also interesting. The initial number of packages maintained by the 216 contributors of Debian 2.0 was 1,101. The corresponding number of packages in Debian 2.1 (around nine months later) for the developers remaining rose to 1,351 and then to 1,457 for Debian 2.2, where the maximum number of packages was achieved. Then it decreased to 1,305 for Debian 3.0, and in the last Debian version it had similar figures as in their first release, although now with half of the maintainers.

This data shows that there has been a continuous increase in the mean number of packages that maintainers are responsible for. While the number of packages per maintainer was slightly above five for the 2.0 release, this number has grown to nine packages per maintainer in release 3.1. It also seems that the mean value keeps on growing, although at a lower pace, and that it has a tendency towards a value around nine as we can see from the last two releases. Given the large amount of time between the last two releases, we can assume that this observed pattern is stable. We have already discussed possible explanations for this behavior with the data about the evolution in number of Debian maintainers (see Table 2).

Figure 3. Half-life of Debian maintainers: How populations shrink over time. The horizontal line at 0.5 shows when a population reaches half-life

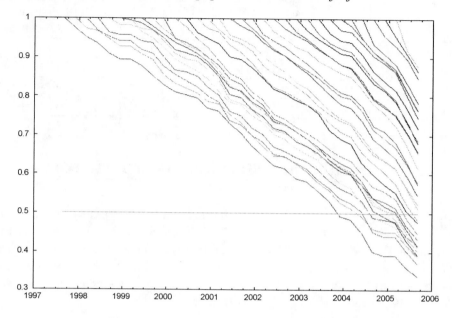

Regarding the involvement of maintainers, we can see from the median that there is a general shift towards maintaining more packages, as the median value starts with three packages and raises up to five for Debian 2.2 and Debian 3.0. The mode, on the other hand, shows that the number of maintainers who only maintain one package decreases over time more quickly than the number of total maintainers (the total number of maintainers drops from 216 to 117, a drop of 46%, while the number of maintainers who maintain only one package decreases from 52 to 20, a 62% drop; this means that the number of maintainers with more than one package shrinks from 164 to 97, which is only 34% less, almost half of the drop for maintainers in charge of a single package). The cause for this may be twofold: On one hand those maintainers could have left the project, and on the other they could have gotten more involved in it by maintaining more packages. In any case, maintaining one package could be seen as a "hot" zone in which nobody stays for a long time and where a decision has to be taken: to get more involved in the project or to leave.

The standard deviation and the Gini coefficient give an idea of the distribution of work. Both values show that there is tendency to have a less equally distributed load of work. Of particular interest is the Gini coefficient, which starts at almost 0.5 and grows up to 0.574. The maximum number of packages that a single maintainer is in charge of grows consequently, from 50 packages in Debian 2.0 to 83 in Debian 3.1. It should be noted that the maximum number of packages of the first three Debian versions under study correspond to a different person than the last two.

Table 3. Packages maintained by the Debian 2.0 maintainers

Date	Release	Maint	Packages	Pkg/Maint	Median	Mode	Std. Dev	Gini	Max
Jul 98	2.0	216	1,101	5.1	3	1 (52)	5.8	0.492	50
Mar 99	2.1	207	1,351	6.5	4	1 (38)	7.3	0.501	55
Aug 00	2.2	188	1,457	7.8	5	1 (33)	9.2	0.515	69
Jul 02	3.0	147	1,305	8.9	5	2 (20)	10.6	0.540	65
Jun 05	3.1	117	1,055	9.0	4	1 (20)	12.1	0.574	83

Investigating Maintainer Experience

In the previous paragraphs we tracked maintainers from Debian version 2.0 over time to see how their contributions evolved. In the following we are going to do the opposite; we will take the last Debian release (Debian 3.1) and will try to track when maintainers first started participating in the project. This will allow us to have a measure of experience in the project. Maintainers who entered in the same release will be grouped and analyzed together.

Figure 4 shows when currently active maintainers got involved in the project. For every maintainer of a package in the latest release, we have investigated in which release their first contribution can be found. In addition to the 117 developers who have made steady contributions since July 1998 (release 2.0), 55 participants got involved before Debian 2.1, and 106 arrived with Debian 2.2. In the last two stable releases, 384 and 652 new maintainers have been identified.

The evolution of the number of packages per maintainer given in Table 4 provides evidence about the impact of experience on the number of packages maintained.

Figure 4. First stable release to which Debian 3.1 maintainers have contributed

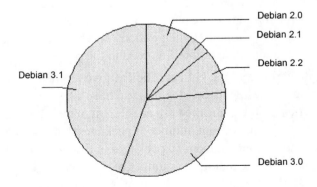

Table 4. First release as maintainer for maintainers in Debian 3.1

Date	Releale	Maint	Packages	Pkg/Maint	Median	Mode	Std. Dev	Gini	Max
Jul 98	2.0	117	1,055	9.0	4	1 (20)	12.1	0.574	83
Mar 99	2.1	55	631	11.5	6	5 (8)	15.1	0.544	81
Aug 00	2.2	106	1,008	8.8	6	1;2 (16)	9.7	0.515	55
Jul 02	3.0	384	2,835	7.2	5	1 (63)	9.5	0.511	121
Jun 05	3.1	652	2,221	4.0	2	1 (246)	7.4	0.570	106

We can see that for the first three versions considered, the values range around 9.0 up to 11.5 packages per maintainer, while in the last two the number of packages is lower. In general, the tendency is that older maintainers have more packages than those who joined later. The exceptions are maintainers who joined with Debian 2.1. For this version, we can see a statistical distortion in the mode as it has a value of 5 while the values for all other versions is 1 (or 1 and 2 in the case of Debian 2.2).

With regard to the median value, we can see that it is also higher for more experienced maintainers, although in this case it is not that clear as we have seen with the mean number of packages. The different behavior of the last two releases is interesting: While Debian 3.1 has a median of two with many (up to 246) maintainers only in charge of one package, those maintainers who entered in Debian 3.0 and who are still active, have a median value of five and a smaller proportion of them maintain a single package. Again, this supports our previous conclusion that maintaining a single package is only a temporary situation.

The standard deviation of the sample does not give us much information in this case. Maybe it stresses the distorted behavior of Debian 2.1 with such a high value; interestingly enough it shows that the data is more homogeneous as we come nearer to Debian 3.1. This is an expected effect, as "younger" maintainers should have a similar (smaller) involvement while "older" ones may vary more. Nonetheless, the Gini coefficient does not necessarily support this finding as the values show no clear tendency over time (the highest value is for Debian 2.0 followed closely by Debian 3.1). This is also the case for the maximum number of packages maintained by a single person which fluctuates from version to version without any predictable direction. In any case, we can see that very active maintainers enter any time in the project, some of them with a surprisingly high involvement. For instance, from July 2002 to June 2005 one new maintainer became in charge of 106 packages! Obviously, the effort needed for the maintenance of a package can vary widely. Further exploration is needed to estimate the effort associated with the maintenance of these packages.

Table 5. Orphaning and adoption of packages

Release 1	Release 2	Orphaned	Adopted	Adopt/Orph	Orph/Total1	Orph/Total2
2.0	2.1	15	14	93.3%	1.3%	1.0%
2.0	2.2	61	40	65.6%	5.5%	1.5%
2.0	3.0	231	171	74.0%	21.0%	4.5%
2.0	3.1	385	253	65.7%	35.0%	4.8%
2.1	2.2	47	31	66.0%	3.0%	1.8%
2.1	3.0	302	220	72.8%	19.4%	5.9%
2.1	3.1	516	332	64.3%	33.1%	6.5%
2.2	3.0	281	207	73.7%	10.8%	5.5%
2.2	3.1	685	433	63.2%	26.3%	8.6%
3.0	3.1	685	435	63.5%	13.4%	8.6%

Note: Each row shows packages present in the older release (first column) and not in the newer ("Orphaned" column), and which of those were adopted. The last columns show the percentages of package "saved" (adopted to orphaned, Adopt/Orph), and orphaned in the newer release to total in the older (Orph/Total1) and newer (Orph/Total2) releases.

Packages of Maintainers who Left the Project

When maintainers leave the project, their packages become unmaintained (the Debian project uses the expression "orphaned"). These packages may be taken up by others ("adopted" in the Debian jargon), or they will not be present in the next stable release. In Table 5 the ratios and numbers of orphaned and adopted packages between any pair of the studied releases are shown.

The table should be read as follows: The first column shows the number of packages, 15, for which their maintainers have left the project (column "Orphaned") and that have been adopted, 14, by other (possibly new) maintainers (column "Adopted") from Debian 2.0 to Debian 2.1. This means that the percentage of orphaned packages that have been adopted is 93.3% (column "Adopt/Orph"), so in this case few packages got lost. The last two columns help situating the amount of orphaned packages we are talking about, giving the share of orphaned packages in comparison to the total number of packages for each release.

Looking at the rest of the rows in the table, we can see that the percentage of adopted packages is very high: more than 60% for all releases considered. This happens even for releases with a very high portion of orphaned packages (for instance, between version 2.0 and 3.1). In other words, even though maintainers who left Debian between July 1997 and June 2005 were responsible for 35.0% of the packages in Debian 2.0, 65.7% of these packages can still be found in version 3.0. We can thus affirm that Debian counts on a natural "regeneration" process for its voluntary contributors and that there is a high probability that the packages of a maintainer who leaves the project will be adopted by others.

Another interesting fact is that the ratio of adopted to orphaned packages is decreasing in later releases. This means that the number of orphaned packages grows more quickly than that of adopted, i.e., there are some packages missing in every new release. If a package is unmaintained and falls off the next release, it will probably not enter a future one. In this study we have only considered removed packages from maintainers who left the project, but it is likely that some software will also be abandoned by maintainers who still remain active and are therefore not covered by this study.

In any case, it should be noted that users are left unsupported when a package (maybe providing a unique functionality) from a previous release is not present in subsequent ones. It may therefore be beneficial to establish mechanisms to ensure that only packages which can be supported in the long term will not be introduced in the first place, or that at least that they be introduced only in a section of the Debian repository which is clearly marked as being less supported.

Experience and Importance

We have used data from the Debian Popularity Contest (presented in detail in the section Debian Popularity Contest) to find out whether more "important' packages are maintained by more experienced volunteers. Table 6 shows the data corresponding to installations and use of packages by developers which are still in the project, and which were already present in the studied releases. In it we can see, for instance, that Debian 2.0 and 3.1 have 117 common maintainers, who are responsible for 1,091 packages which have been installed 1,305,907 times and 576,991 that are regularly used.

Table 6. Installations and regular use of packages

Release	CMaint	CPkg	Installations	Votes	Inst/Maint	Votes/Maint
2.0	121	1,091	1,305,907	576,991	10792.6	4768.5
2.1	176	1,722	1,584,413	673,236	9002.3	3825.2
2.2	290	2,730	2,217,199	885,448	7645.5	3053.3
3.0	683	5,565	3,923,753	1,405,322	5744.9	2057.6
3.1	1315	7,989	5,248,869	1,711,496	3991.5	1301.5

The CMaint column shows how many maintainers Debian 3.1 had in common with the release in the first column, while the CPkg shows the number of packages maintained by them. Columns Installations and Votes give the sum of the packages installed and voted (used regularly) for those packages maintained by common maintainers. The last two columns show the ratios of both to common maintainers.

If we take the number of installations per maintainer and the number of regularly used packages per maintainer ("Votes/Maint') we can answer the question we proposed in the section Research Objectives and Goals. According to our hypothesis, these ratios decrease over time, which would mean that more experienced volunteers maintain packages which are installed and used more often. In fact, this can be observed through all Debian releases. An alternative (or complementary) explanation to our initial hypothesis is that many of the essential components of the Debian system were introduced in the first releases, and that new packages are mostly add-ons and software that are not installed and used that often.

Conclusion and Further Work

We have conducted a quantitative study of the evolution of the Debian maintainer-ship over the last six-and-a-half years. We have retrieved and analyzed publicly available data in order to find out how Debian handles the volatility of the volunteers who made it happen.

Some of the most interesting findings are:

- Both the number of Debian maintainers and the number of packages per maintainer grow over time.
- The number of maintainers from previous releases who remain active is very high, with an estimated half-life of around 6.5 years (78 months). Slightly less than half of the maintainers from Debian 2.0 still contribute to the current release after more than 7 years.
- Developers tend to maintain more and more packages as they gain experience in the project.
- However, this does not mean that maintainers who have been in the project for more time maintain more packages than newer maintainers. In fact, in the latest release, the highest packages per maintainer ratio is shown by those entering the project around the year 2000.

From these facts, it can be said that Debian maintainers tend to commit to the project for long periods of time. However, there is a worrisome trend towards a higher and higher ratio of packages per maintainer, which could imply scalability problems as the number of packages in the distribution increases, if the project doesn't admit a proportional number of developers.

Another issue on which we have focused is what happens to those packages that were maintained by developers who left the project. Most of them are taken over by other maintainers, so that we can state that a natural "regeneration" exists. Based on the data we have researched, those packages which are not adopted by other maintainers in the next release, and are therefore not present in it, are unlikely to be re-introduced in future releases.

Finally, we have also found that more experienced maintainers are responsible for packages which are installed more often and used more regularly.

In addition to the new insights gained in this investigation, we have proposed a number of further studies to elaborate on the findings of the present chapter. In particular, team maintenance and its impact on the quality of packages would be interesting to research. It is also not clear why there is an increase in the ratio of packages per maintainer. Possible explanations are that better tools and practices lead to more efficiency, or that with the success of libre software, new volunteers show more motivation and commitment, but more data is needed before these explanations can be conclusive.

From a more general point of view, this study explores the behavior of volunteers in libre software projects and provides some answers as to why these kinds of voluntary contributions are capable of producing such large, mature and stable systems over time, even when the project has no means for forcing any single developer to do any given task and when members may leave the project during important development phases. It is impossible to infer the behavior of volunteer developers just from the study of a single project, but given the size and relevance of the Debian project, at least some conclusions can be exposed as hypotheses for validating in later research efforts.

One of them is the stability of volunteer work over time. The mean life of contributors in the project is probably longer than in many software companies, which would have a clear impact on the maintenance of the software (it is likely that developers with experience in a module are available for its maintenance over long periods of time). Another one is that volunteers tend to take over more work with the passing of time if they remain in the project: In other words, they voluntarily increase their responsibilities in the project. Whether this is because it is easier for them because of their experience, or because they devote more effort to the project, is for now an open question. Yet a third one is the stability of the voluntary effort when some individuals leave the project: Most of their work is taken over by other developers. Therefore, despite being completely based on volunteers, the project organizes itself rather well with respect to drop-outs, which is an interesting lesson about how the project can survive in the long term.

As a final summary, we have found that given that there are no formal ways of forcing a developer to assume any given task, voluntary efforts seem to be more stable over time, and more reliable with respect to individuals leaving the project than we had initially expected.

Acknowledgment

The work of Martin Michlmayr has been funded in part by Google, Intel, and the EPSRC. The work of Gregorio Robles and Jesus M. Gonzalez-Barahona has been funded in part by the European Commission under the CALIBRE CA, IST program, contract number 004337. We would also like to thank the anonymous reviewers for their extensive comments.

References

Crowston, K., & Howison, J. (2005). The social structure of free and open source software development. *First Monday, 10*(2).

Debian Social Contract. (2006). Debian social contract. Retrieved from http://www. debian.org/social_contract

Dinh-Trong, T. T., & Bieman, J. M. (2005). The FreeBSD project: A replication case study of Open Source development. *IEEE Transactions on Software Engineering, 31*(6), 481-494.

Garzarelli, G., & Galoppini, R. (2003). *Capability coordination in modular organization: Voluntary FS/OSS production and the case of Debian GNU/Linux.*

German, D. (2004). The GNOME project: A case study of open source, global software development. *Journal of Software Process: Improvement and Practice, 8*(4), 201-215.

Ghosh, R. A., Glott, R., Krieger, B., & Robles, G. (2002). Survey of developers (Free/libre and open source software: Survey and study). Technical report, International Institute of Infonomics, University of Maastricht, The Netherlands. Retrieved from http://www.infonomics.nl/FLOSS/report

Ghosh, R. A., & Prakash, V. V. (2000). The orbiten free software survey. *First Monday, 5*(7). Retrieved from http://www.firstmonday.dk/issues/issue5_7/ghosh/

Gonzalez-Barahona, J. M., Ortuno Perez, M. A., de las Heras Quiros, P., Centeno Gonzalez, J., & Matellan Olivera, V. (2001). Counting potatoes: The size of Debian 2.2. *Upgrade Magazine, II*(6), 60-66.

Gonzalez-Barahona, J. M., Robles, G., Ortuno Perez, M., Rodero-Merino, L., Centeno Gonzalez, J., et al., (2004). Analyzing the anatomy of GNU/Linux distributions: Methodology and case studies (Red Hat and Debian). In S. Koch (Ed.), *Free/open source software development* (pp. 27-58). Hershey, PA: Idea Group Publishing.

Herbsleb, J. D., Mockus, A., Finholt, T. A., & Grinter, R. E. (2001). An empirical study of global software development: Distance and speed. In *ICSE '01: Proceedings of the 23rd International Conference on Software Engineering* (pp. 81-90).

Hertel, G., Niedner, S., & Herrmann, S. (2003). Motivation of software developers in open source projects: An Internet-based survey of contributors to the Linux kernel. *Research Policy, 32*(7), 1159-1177.

Hunt, F., & Johnson, P. (2002). On the Pareto distribution of open source projects. In *Proceedings of Open Source Software Development Workshop*, Newcastle, UK.

Koch, S., & Schneider, G. (2002). Effort, cooperation and coordination in an open source software project: GNOME. *Information Systems Journal, 12*(1), 27-42.

Lameter, C. (2002). *Debian GNU/Linux: The past, the present and the future.* Retrieved from http://telemetrybox.org/tokyo/

Michlmayr, M. (2004). Managing volunteer activity in free software projects. In *Proceedings of the USENIX 2004 Annual Technical Conference, FREENIX Track*, Boston (pp. 93-102).

Michlmayr, M., & Hill, B. M. (2003). Quality and the reliance on individuals in free software projects. In *Proceedings of the 3rd Workshop on Open Source Software Engineering*, Portland, OR (pp. 105-109).

Mockus, A., Fielding, R. T., & Herbsleb, J. D. (2002). Two case studies of open source software development: Apache and Mozilla. *ACM Transactions on Software Engineering and Methodology, 11*(3), 309-346.

Monga, M. (2004). From bazaar to kibbutz: How freedom deals with coherence in the Debian project. In *Proceedings of the 4th Workshop on Open Source Software Engineering*, Edinburg, Scotland, UK.

O'Mahony, S. (2003). Guarding the commons: How community managed software projects to protect their work. *Research Policy, 32*, 1179-1198.

Robles, G. (2006). *Empirical software engineering research on libre software: Data sources, methodologies and results.* PhD thesis, Universidad Rey Juan Carlos.

Robles, G., Gonzalez-Barahona, J. M., & Michlmayr, M. (2005). Evolution of volunteer participation in libre software projects: Evidence from Debian. In

Proceedings of the 1ˢᵗ International Conference on Open Source Systems, Genoa, Italy (pp. 100-107).

Endnotes

[1]   In this chapter we will use the term "libre software" to refer to any software licensed under terms compliant with the Free Software Foundation definition of "free software," and the Open Source Initiative definition of "open source software," thus avoiding the controversy between those two terms.

[2]   http://popcon.debian.org

[3]   All the code used has been released as libre software, and can be obtained from http://libresoft.dat. escet.urjc.es/index.php?menu=Tools

[4]   The Gini coefficient is a normalized measure of inequality; values near 0 point out equal distributions while values close to 1 are indicative for high inequalities.

Chapter II

Measuring the Efficiency of Free and Open Source Software Projects Using Data Envelopment Analysis

Stefan Koch, Vienna University of Economics and
Business Administration, Austria

Abstract

In this chapter, we propose for the first time a method to compare the efficiency of free and open source projects, based on the data envelopment analysis (DEA) methodology. DEA offers several advantages in this context, as it is a non-parametric optimization method without any need for the user to define any relations between different factors or a production function, can account for economies or diseconomies of scale, and is able to deal with multi-input, multi-output systems in which the factors have different scales. Using a data set of 43 large F/OS projects retrieved from SourceForge.net, we demonstrate the application of DEA, and show that DEA indeed is usable for comparing the efficiency of projects. We will also show additional analyses based on the results, exploring whether the inequality in work distribution within the projects, the licensing schem,e or the intended audience have an effect on their efficiency. As this is a first attempt at using this method for F/OS projects, several future research directions are possible. These include additional work on determining input and output factors, comparisons within application areas, and comparison to commercial or mixed-mode development projects.

Introduction

In the last years, free and open source software (also sometimes termed libre software) has gathered increasing interest, both from the business and academic world. As some projects in different application domains like, most notably, the operating system Linux together with the suite of GNU utilities, the office suites GNOME and KDE, Apache, sendmail, bind, and several programming languages have achieved huge success in their respective markets, both the adoption by commercial companies, and also the development of new business models by corporations both small and large like Netscape or IBM, have increased.

Currently, any comparison of free and open source (F/OS) software projects is very difficult. There is increased discussion on how the success of F/OS projects can be defined (Stewart, 2004; Stewart & Ammeter, 2002; Crowston et al., 2004; Crowston et al., 2003), using, for example, search engine results as proxies (Weiss, 2005). In addition, the process applied in these projects can differ significantly.

In this chapter, we propose to compare F/OS projects according to their efficiency in transforming inputs into outputs. For any production process, this efficiency and productivity are key indicators in comparison to other processes. For F/OS projects, two levels of analysis are of interest: The F/OS process in general is different than commercial software development processes, and the process variance between F/OS projects is also high. In both cases, a main difference, and a main argument for adopting a process or elements from it, is the efficiency. Neither a commercial enterprise nor an F/OS project would knowingly and willingly waste scarce resources by using an inefficient development process. This necessitates one to compute and compare the efficiency of F/OS projects to gain an understanding of the results any process decision has on the outputs, which could lead to identifying best practices, and thus increasing the overall efficiency and output of all projects.

To this end, we propose to apply the method of data envelopment analysis (DEA), which is a non-parametric optimization method for efficiency comparisons without any need for the user to define any relations between different factors or a production function. In addition, DEA can account for economies or diseconomies of scale, and is able to deal with multi-input, multi-output systems in which the factors have different scales. Efficiency and productivity in software development is most often denoted by the relation of an effort measure to an output measure, using either lines-of-code (Park, 1992) or, preferably due to independence from programming language, function points (Albrecht & Gaffney, 1983). While this approach can be problematic in an environment of commercial software development as well, due to missing components, especially of the output (e.g., Kitchenham & Mendes, 2004, agree that productivity measures need to be based on multiple size measures), there are additional problems in the context of F/OS development which point towards DEA as an appropriate method.

In F/OS projects, normally the effort invested is unknown, and therefore might need to be estimated (Koch, 2004, 2005); they are also more diverse than commercial projects, as they include core team members, committers, bug reporters and several other groups with varying intensity of participation. Besides that, the outputs also can be more diverse. In the general case, the inputs of an F/OS project can encompass a set of metrics, especially concerned with the participants. So, in the simplest case, the number of programmers and other participants can be used. The output of a project can be measured using several software metrics, as most easily, the number of LOC, files, checkins to the source code control system, postings, bug reports, characteristics of development speed (e.g., coefficients of a software evolution equation estimated) or even metrics for product attributes like McCabe's cyclomatic complexity (McCabe, 1976) or object-oriented metrics (e.g., the Chidamber-Kemerer suite, Chidamber & Kemerer, 1994). This range of metrics both for inputs, outputs, and their different scales, necessitates application of an appropriate method, which DEA can be.

The main result of applying DEA for a set of projects is an efficiency score for each project. This score can serve different purposes as mentioned above: First, single projects can be compared accordingly, but also groups of projects, for example those following similar process models, located in different application domains, or simply of different scale can be compared to determine whether any of these characteristics lead to higher efficiency. Lastly, this method could also be used for a comparison with traditional, commercial, or even hybrid projects, which in addition to volunteer resources, also feature monetary input from an organization. In this way, important questions concerning software development in general, and especially in F/OS projects, can be answered: Which methods allow for the most efficient application of the available manpower? Is F/OS software development a new, more efficient way of producing software? Are projects on a larger scale more efficient than those on smaller scale? Is a high concentration of work on a small number of heads efficient?

In this chapter, we will demonstrate a first application of DEA to compare the efficiency of a set of F/OS projects. After a short introduction to F/OS, we will describe the method of DEA in general. Based on this, the application will be detailed, starting with a description of the methodology of data retrieval and the respective data set to be used. We will then discuss how to set up the DEA model in this case, taking into account the available data, and give the results. In a section on future research, possible enhancements to this work will be given, in addition to applications of the results obtained.

Free and Open Source Software

For a quick definition of free and open source software, both terms needs to be analyzed (Laurent, 2004; Dixon, 2003; Rosen, 2004). The term open source as used by the Open Source Initiative (OSI) is defined using the Open Source Definition (Perens, 1999), which lists a number of rights which a license has to grant in order to constitute an open source license. These include, most notably, free redistribution, inclusion of source code, to allow for derived works which can be redistributed under the same license, and integrity of author's source code. The Free Software Foundation (FSF) advocates the term "free software" (Stallman, 2002). A software is defined as free if the user has the freedom to run the program, for any purpose, to study how the program works and adapt it to his or her needs, to redistribute copies and improve the program, and release these improvements to the public. Access to the source code is a necessary precondition. In this definition, open source and free software are largely interchangeable. The GNU project itself prefers copylefted software, which is free software whose distribution terms do not let redistributors add any additional restrictions when they redistribute or modify the software. This means that every copy of the software, even if it has been modified, must be free software, a prescription embodied in the most well-known and important license, the GNU General Public License (GPL).

Not only is F/OS software unique in its licenses and legal implications, but also in its development process. The main ideas of this development model are described in the seminal work of Raymond (1999), *The Cathedral and the Bazaar*, in which he contrasts the traditional type of software development of a few people planning a cathedral in splendid isolation with the new collaborative bazaar form of open source software development. In this, a large number of developer-turned-users come together without monetary compensation (Raymond, 1999; Hertel et al., 2003) to cooperate under a model of rigorous peer review and take advantage of parallel debugging that leads to innovation and rapid advancement in developing and evolving software products. In order to allow for this to happen and to minimize duplicated work, the source code of the software needs to be accessible which necessitates suitable licenses, and new versions need to be released often. Today, agile methods like eXtreme Programming or the strict release processes in place in several open source projects (Holck & Jorgensen, 2004) give evidence to mixed forms of development. Currently, empirical research on similarities and dissimilarities between F/OS development and other development models is still proceeding (Mockus et al., 2002; Koch, 2004).

Data Envelopment Analysis

Production and Efficiency

The production of a good or service can be formalized using a production function, which gives the maximum possible output for a given amount of input (Varian, 2005). Several mathematical forms for production functions have been used in literature, including a linear form or a Cobb-Douglas production function, necessitating an appropriate number of parameters for defining the function in a concrete case. The notion of returns to scale is intimately linked to this concept: If the inputs are scaled by a constant factor, for example, twice as much input is used, constant returns to scale would mean that the output is scaled by the same amount. If increasing returns to scale are present, the output would increase more than that, less with decreasing returns to scale (Varian, 2005). Of course, the returns to scale might be different at different levels of production, leading to an ideal size of production (or firm). The term of economies of scale refers to the decreased per unit cost as output increases. Both the production function to be used (Hu, 1997), and the returns to scale (Banker & Kemerer, 1989; Banker & Slaughter, 1997) have long been a topic of discussion in software development.

For comparing the efficiency of different firms (or units), the most basic approach is to compute the ratio of an output to an input. Those firms (or units) exhibiting the highest ratio are able to produce the most output given an input, thus are most efficient. For the area of software development, efficiency or productivity is most often denoted as lines-of-code, or preferably function points (Albrecht & Gaffney, 1983) due to their technology neutrality, per person-month of effort. Problems with this approach are that it is computationally not usable for multi-input, multi-output situations, and that it assumes constant returns to scale.

Main Concepts of DEA

The principle of the border production function was introduced by Farell (1957) for measuring technical efficiency and enhanced by Charnes, Cooper, and Rhodes (1978a) into the first data envelopment analysis model (the CCR model). The term DEA is used for the first time in an evaluation of school programs for the support of handicapped children. The object of analysis the DEA considers is very generally termed "decision making unit" (DMU). This term includes with relative flexibility each unit which is responsible for the transformation of inputs into outputs. As different applications show, this definition therefore covers hospitals, supermarkets, schools, bank branches, and others.

The basic principle of DEA can be understood as a generalization of the normal efficiency evaluation as described above by means of the relationship from an output to an input into the general case of a multi-output, multi-input system without any given conversion rates or equal units for all factors. For the area of software development, this can be understood as adding, for example, pages of documentation to lines-of-code and function points as an output factor. In contrast to other approaches, which require the parametric specification of a production function, DEA measures production behavior directly and uses this data for the evaluation of all DMUs. The DEA derives a production function from mean relations between inputs and outputs (whereby it is only assumed that the relation is monotonous and concave), by determining the outside cover of all production relations, while, for example, a regression analysis estimates a straight line through the centre of all production relations. The DEA identifies "best practicing" DMUs, which lie on the production border. Thus, any outliers and measuring and/or data errors may exert a strong influence on the results, and therefore special attention is to be given to sensitivity analyses. In order to determine the production function, the efficient DMUs are linked in sections with one another. A DMU is understood as being efficient if none of the outputs can be increased without either or several of the inputs increasing or other outputs being reduced, as well as when none of the inputs can be reduced without reducing either one or more outputs or increasing other inputs.

Computation

For each DMU, an individual weighting procedure is used over all inputs and outputs. These form a weighted positive linear combination, whereby the weights are specified in such a way that they maximize the production relationship of the examined unit in order to let these become as efficient as possible. The efficiency of an examined unit is limited with 1. That means it that no apriori weightings are made by the user, and that the weights between the DMUs can be different. However, the definition of the relevant inputs and outputs is necessary. For each evaluation object, the DEA supplies a solution vector of weighting factors and a DEA efficiency score. If this score is equal to 1, then the DMU is DEA efficient. In this context, DEA efficiency means that within the selected model variant no weighting vector could be found which would have led to a higher efficiency value. All those DMUs which are not clearly DEA inefficient compared with the others are thus DEA efficient. Any inefficiency can therefore not be ruled out completely. For inefficient DMUs, weighting factors were found in the context of the selected model variant which resulted in a higher efficiency value in the case of at least another DMU.

For each inefficient DMU, the DEA returns a set of efficient DMUs which exhibit a similar input/output structure and lie on the production border near to the inefficient DMU (this is also termed a reference set or DEA benchmark). Using this information, an idea of which direction an increase in efficiency is possible can be gained. Because the relative efficiency measure is based on the distances from actually existing DMUs and is thus easily comprehensible, DEA is better suitable as an analysis tool than other methods.

Models of DEA

The first model of the DEA was introduced by Charnes, Cooper, and Rhodes (1978b) and is designated with the initial letters of their surnames as the CCR model. They pose four assumptions for the production possibility set, which are convexity (each linear combination of two DMUs results in a production possibility within the valid range), possibility for inefficient production (all production possibilities which are more inefficient than the known DMUs are permissible), constant returns to scale, and minimum extrapolation (the production possibility range covers all observed DMUs). The CCR model measures the efficiency of each DMU by examining the relationship of the outputs to the inputs, optimally weighted for the DMU. This optimization happens under the constraint that equivalent conditions for no other DMU result in a value exceeding 1. By this maximization each DMU receives the optimum weighting given the constraint.

The different basic models of the DEA can be divided on the basis of two criteria: On the one hand, the orientation of the model, on the other, the underlying assumption regarding the returns to scale of the production process. With input-oriented models, the reduction of the input vector maximally possible with the given manufacturing technology is determined, whereas with output-oriented models the maximally possible proportional increase of the output vector is determined. Input orientation is generally present if an enterprise function (for example manufacturing) is to be analyzed, which minimizes the resources consumption, but can only affect the output in a limited way. The returns to scale can be assumed either as being constant as in the CCR model described above, or variable. With constant returns to scale, size-induced productivity differences are considered in the efficiency evaluation, with variable returns to scale, the differences are neutralized by the model. The most common example of a model with variable returns to scale is an advancement to the CCR model by Banker, Charnes, and Cooper (1984), the BCC model. This model includes an additional measuring variable in the fundamental equation to capture rising, falling, or constant returns to scale.

Applications

DEA models are generally used where scarce resources are to be used in a goal-oriented environment. The first DEA models were developed to measure the efficiency of non-profit units, for whose inputs and outputs no clear market prices exist and also no clear evaluation relations are present otherwise. The entire public sector, therefore, represents a typical area of application. Emphases of the first investigations were therefore health service, in particular hospitals, as well as school advancement programs (Charnes et al., 1978a). In further consequence, the DEA was also applied to the evaluation of the efficiency of private business units as, for instance, branches of banks or supermarkets, especially if the efficiency is not only to be measured in profit, but also by parameters, as for environmental conditions, down-times, or customer satisfaction.

In the area of software development, DEA was only rarely applied. Banker and Kemerer (1989) use this approach in order to prove the existence of both rising and falling returns to scale. This approach first arises in small (the others in larger) software development projects. Based on published collections of project data records, the authors compute in each case the point of the maximum productivity (most productive scale size) within the set of projects, starting from which the returns to scale begin to fall. Banker and Slaughter (1997) use the DEA in the area of maintenance and enhancement projects. It can be proven that rising returns to scale are present, which would have made a cost reduction of around 36 percent possible when utilized. This was prevented, however, by organizational constraints—for example, high punishments for exceeding the completion date. Mayrhauser et al. (2000) also report applying DEA on a data set consisting of 46 software projects from the NASA-SEL database to analyze objective variables and their impact on efficiency. They suggest combining this with the results of applying principal component analysis (PCA) in an analysis of subjective variables. An investigation of ERP projects was done by Myrtveit and Stensrud (1999). They used 30 SAP R/3-projects of a consulting firm for the application of the DEA. Kitchenham (2002) gives an in-depth discussion on the application of DEA in software development.

Application of DEA for Free and Open Source Software Projects

Methodology

For performing the proposed efficiency comparison of F/OS software projects, the information contained in software development repositories will be used. These repositories contain a plethora of information on the underlying software and the associated development processes (Cook et al., 1998; Atkins et al., 1999). Studying software systems and development processes using these sources of data offers several advantages (Cook et al., 1998): This approach is very cost-effective, as no additional instrumentation is necessary, and it does not influence the software process under consideration. In addition, longitudinal data are available, allowing for analyses considering the whole project history.

Depending on the tools used in a project, possible repositories available for analysis include source code versioning systems, bug reporting systems, or mailing lists. Many of these have already been used as information sources for closed source software development projects. For example, Cook et al. (1998) present a case study to illustrate their proposed methodology of analyzing in-place software processes. They describe an update process for large telecommunications software, analyzing several instances of this process using event data from customer request databases, source code control, modification request tracking databases and inspection information databases. Atkins et al. (1999) use data from a version control system in order to quantify the impact of a software tool, a version-sensitive editor, on developer effort.

In open source software development projects, repositories in several forms are also in use, in fact, from the most important communication and coordination channels, as the participants in any project are not collocated. Therefore, only a small amount of information cannot be captured by repository analyses because it is transmitted inter-personally. As a side effect, the repositories in use must be available openly and publicly, in order to enable as many persons as possible to access them and to participate in the project. Therefore, open source software development repositories form an optimal data source for studying the associated type of software development.

Given this situation, repository data have already been used in research on open source software development. This includes in-depth analyses of small numbers of successful projects like Apache and Mozilla (Mockus et al., 2002) and GNOME (Koch & Schneider, 2002), using mostly information provided by version-control-systems, but sometimes in combination with other repository data, like from mailing list archives. Large-scale quantitative investigations spanning several projects going into software development issues are not yet as common, and have mostly been limited to using aggregated data provided by software project repositories (Crowston & Scozzi, 2002; Hunt & Johnson, 2002; Krishnamurthy, 2002), meta-information included in Linux Software Map entries (Dempsey et al., 2002), or data retrieved directly from the source code itself (Ghosh & Prakash, 2000).

Data Collection and Data Set

For this efficiency comparison, a data set covering several projects was needed. Therefore, we used a subset of a data set already available from prior research derived from SourceForge.net, the software development and hosting site. The mission of SourceForge.net is "to enrich the open source community by providing a centralized place for open source developers to control and manage open source software development." To fulfill this mission goal, a variety of services is offered to hosted projects, including tools for managing support, mailing lists and discussion forums, Web server space, shell services and compile farm, and source code control. While SourceForge.net publishes several statistics, e.g., on activity in their hosted projects, this information was not detailed enough for the proposed analysis. For example, Crowston and Scozzi (2002) used the available data for validating a theory for competency rallying, which suggests factors important for the success of a project. Hunt and Johnson (2002) have analyzed the number of downloads of projects occurring, and Krishnamurthy (2002) used the available data of the 100 most active mature projects for an analysis.

The data collection method utilized is described in detail in Hahsler and Koch (2005). It is based on automatically extracting data from the Web pages, and especially the source code control system, in the form of CVS, and subsequently parsing the results and storing the relevant results, for example, size, date, and programmer of each checkin, in a database. Most of this work was done using Perl scripts. This resulted in a number of 8,621 projects for which all relevant information could be retrieved. More detailed analyses of these data can be found in Koch (2004). For this research, we chose to use the subset of largest projects, in order to have a limited set exhibiting

Table 1. Descriptive statistics of data set (n=43)

	Min.	Max.	Mean	Std. Dev.
Size (LOC)	533321,00	7669643,00	2014731,09	2124643,99
Total of LOC added	541148,00	10773710,00	2478442,26	2447076,77
Total of LOC deleted	7827,00	3104067,00	463711,16	571624,21
Number of Files	904,00	23594,00	6381,79	5683,65
Start date	1990/08/09	2001/04/17	1999/01/16	n/a
Checkins	1897,00	133759,00	27485,40	26986,30
Total number of programmers	5,00	88,00	18,35	15,62
Mean number of programmers per month	1,40	13,00	4,39	2,86
Cumulated active programmers (person-years)	1,00	55,40	9,78	9,79
Number of months	5,00	133,00	27,44	22,12
Status	,00	6,00	4,21	1,55
Inequality (based on checkins)	,05	,87	,33	,24

similar characteristics. We therefore defined a threshold of at least five developers and 500,000 lines-of-code, which resulted in 43 projects. Table 1 gives descriptive statistics for these projects. While most of the variables are self-descriptive, a few notes are in order: The total number of programmers gives all different persons who have contributed code over the complete lifetime of the project. In contrast, a programmer is defined to be active in a month if he showed activity during this time interval. The mean number of programmers per month is based on this definition, as are the cumulated active programmers. This last number can be seen as an effort indicator, denoting the sum of the active programmers for each month over the total lifetime. It is also given in person-years, although an F/OS programmer active in a given month will not necessarily have worked 40 hours per week. For example, Hertel et al. (2003) report that in the developer group of Linux kernel contributors participating in their survey, about 18.4 hours per week are spent on open source development by each person. This number could be used to convert the effort to "real" person-years, but, as only F/OS projects are to be compared here, this is not done. The status is an indicator assigned by the project administrator within SourceForge.net and aims at reflecting the phase of a project in the development lifecycle. It has seven possible values, reaching from planning, pre-alpha, alpha, beta to production/stable and mature, and to inactive. The last variable is used to depict the inequality of work distribution within the development team (Robles et al., 2004), with values nearer to 1 showing higher inequality.

DEA Model Definition

For computing DEA models, different software products are available, some of which are freeware. In this case, the program accompanying the book by Cooper, Seiford, and Tone (2000) was used, which can solve different DEA models, input or output-oriented, as well as with constant or variable returns to scale.

The first choices to be taken concern the definition of input and output factors, as well as the model to be applied. Banker and Kemerer (1989) have demonstrated the existence of both increasing and decreasing returns to scale in software projects; Myrtveit and Stensrud (1999) also recommend to use a model with variable returns to scale. Using a data set of maintenance and extension projects, Banker and Slaughter (1997) have found increasing returns to scale. Kitchenham (2002) gives an overview of research results and reasons for differences on economies and diseconomies of scale in software development. Regarding the orientation of the model, an output-orientation might seem more appropriate. Given a certain input which can be acquired, that is, participants attracted, the output is to be maximized. According to this reasoning, the BCC-O model is applied.

Regarding the definition which factors are to be used as inputs and outputs, it is to be considered that with an increase in the number of factors, more DMUs, that is, projects, are estimated to be efficient, in particular if the database is relatively small. Also, the availability of factors in the data set limits the possibilities. In this case, we selected to use the total number of programmers and their effort as inputs, and size, LOC added and deleted, files, checkins and status as output factors. Naturally, this selection is based on the available data, and could be changed. This point will be discussed in more detail within the section on future research.

Results

The results of a DEA analysis provide for each DMU, in this case F/OS project, an efficiency measure as defined above, that is, of value 1 if no possibilities for improvements in transforming inputs into outputs have been detected given the data. As the DEA model set up in this chapter is output-oriented, any other value below 1 means that a proportional increase of outputs seems possible given the inputs of a project. A DEA efficiency of 0,75 would translate into a possible output increase of 25% for this project. In addition, a reference set is reported, which contains DMUs with similar input-output relations and lie on the production border near to the inefficient DMU. Table 2 gives an extract of the DEA results on project level from the data set, showing two different projects, one of which is deemed DEA efficient. As can be seen, project number 27, mysql, is DEA efficient, while number 3, berlin, has a lower efficiency score of 0,518, which translates into a possible respective increase in outputs. The reference set reported contains the efficient projects mysql, gkernel and python, which also have weights assigned relating to their relative importance in showing the inefficiency of berlin. In this case, mysql is the most important project, which means that it is most similar in input-output relation, and orienting at this project would show the most promising path for efficiency increase.

From all DEA efficiency scores, an efficiency ranking over all projects can be derived, which is given in Table 3 for the data set used here. As can be seen, several projects are deemed DEA efficient, therefore have a value of 1 and are ranked at the top. This list gives two major informations per project: Their rank is a measure

Table 2. Results from DEA on project level (extract)

No.	DMU	Score	Reference set					
...								
3	berlin	0,518	gkernel	0,211	mysql	0,597	python	0,192
...								
27	mysql	1,000	mysql	0,999				
...								

Table 3. Ranking of projects according to DEA efficiency

Rank	DMU	Score
1	Aaf, crossfire, gkernel, lpr, musickit, mysql, nfs, python, u4x, winex	1,000
11	ceps, mesa3d, tcl	0,999
14	linuxsh	0,995
15	linux-apus	0,963
16	Xfce	0,954
17	linux-mac68k	0,903
18	jboss	0,896
19	firebird	0,878
20	cvsgui	0,866
21	enlightenment	0,860
22	quake	0,852
23	sdcc	0,852
24	zangband	0,847
25	opensta	0,840
26	lesstif	0,837
27	ghostscript	0,835
28	qpe	0,833
28	htdig	0,833
30	wf-gdk	0,829
31	dri	0,823
32	crystal	0,769
33	lsb	0,742
34	acs-pg	0,701
35	mahogany	0,691
36	squid	0,670
37	omseek	0,671
38	clocc	0,667
39	linux-vr	0,574
40	berlin	0,518
41	openquartz	0,433
42	cmusphinx	0,369
43	webbase	0,085

Table 4. Overview of DEA results for data set

No. of DMUs	43
Average	0,828
Std. Dev.	0,196
Maximum	1,000
Minimum	0,085
Number of DEA-efficient DMUs	10
Frequency in Reference Set	
Peer set	Frequency to other DMUs
aaf	4
crossfire	8
gkernel	24
lpr	3
musickit	3
mysql	28
nfs	2
python	18
u4x	1
winex	0

of the standing within the total group, while the efficiency score gives an indication of the amount of output increase possible.

For an overview of the results, see also Table 4. In this table, statistics on the efficiency scores in the total population are given. Overall, 10 different projects have been classified as DEA efficient (which is also visible from Table 3), the mean efficiency score with 0,828 seems relatively high. For each efficient project, the number of times it appears in the reference sets of non-efficient projects is also given. This can be used as an indicator of the relative importance of this project in determining efficiency scores, and would also be important for performing sensitivity analyses.

These results, besides demonstrating that differences in efficiency between F/OS projects do indeed exist, can then be used in several ways: First, for each single project, both an easy-to-interpret examination of the current status is given, and an indication for possible improvements in form of the reference set of similar but efficient projects is offered. Secondly, and more important, the results allow for additional analyses beyond single projects, but spanning groups of projects. Using measures of centrality like the mean or median efficiency score, groups of projects can be compared. The groups to be considered can be based on different characteristics available, including software process, application domains, scale, or even license. Using any of the criteria, the population can be divided in two or more groups, and the distribution of efficiency scores can be compared statistically, using the hypothesis that the dividing variable has an effect on efficiency. If any statistically significant difference in efficiency shows up, the hypothesis is validated, meaning that indeed projects showing a certain characteristic tend to be more (or less) efficient.

In this chapter, first an example is detailed based on comparing different software process designs, using the inequality of contributions within projects as a dividing factor. We will therefore compute a correlation coefficient between the inequality measure as given above and the efficiency score of projects. In this case, the underlying hypothesis, that a relationship between group working style as depicted by the resulting inequality and efficiency exists, can not be confirmed: There is no significant correlation between both measures, which shows that a higher inequality does not have any effect on efficiency.

As a second example, the hypothesis that a very stringent copyleft-licensing scheme increases the productivity of the participants is explored. We therefore divide the set of projects into two groups, one consisting of projects licensed under GNU GPL only, resulting in 18 projects out of 43, the other with all remaining projects. Then, a non-parametric Mann-Whitney U-test is employed to test the distribution of the efficiency scores between the groups. In this case, there is no significant difference, so the hypothesis that a copyleft-licensing scheme has an impact on efficiency has to be discarded.

As a last example, we base the grouping of projects on the intended audience, with one group targeting developers and system administrators only, resulting in 20 projects, and others, to explore whether this characteristic has any impact on efficiency. Also in this case, there is no significant difference in efficiency between the two groups, again showing that the intended audience does not lead to a change in efficiency.

Future Research

As this has been a first attempt of applying DEA to the context of F/OS projects, numerous directions for future research exist. The first area of research identified with the data set in this chapter is the definition of relevant input and output factors. We propose to add at least two additional factors, the first being contributors other than programmers as an input factor, the second being a mix of different product metrics to capture aspects of the software system produced. This could include concepts like McCabe's cyclomatic complexity (McCabe, 1976) or object-oriented metrics, for example, the Chidamber-Kemerer suite (Chidamber & Kemerer, 1994).

Also other outputs could be measured and included, for example postings to mailing lists, Web site metrics, and even popularity measures (Weiss, 2005).

In this chapter, we have compared a set of relatively large F/OS projects. It would also be interesting to include smaller projects, in which case the issue of variable or constant returns to scale would increase in importance. A comparison might also be useful to be done within a certain application area only, for example frameworks for Web service implementation, to eliminate additional influences.

Of maybe even more interest than those considerations given above on setting up a DEA might be analyses based on the respective results. After having computed efficiency scores for a set of projects, other characteristics can be used in order to explore reasons for any differences in efficiency. Possible characteristics would include mostly aspects of the software development process employed, similar to those in this chapter, which used the inequality within the development team as an explanatory variable. Respective comparisons could also be extended to include projects following very different approaches, like commercial projects employing agile methodologies, or projects with mixed participation from volunteers and paid contributors. To allow for these comparisons, additional characteristics of the projects considered need to be collected.

Conclusion

In this chapter, we have proposed for the first time a method to compare the efficiency of F/OS projects. The method used is the DEA, which is well-established in other fields and offers several advantages in this context as well: It is a non-parametric optimization method without any need for the user to define any relations between different factors or a production function, it can account for economies or diseconomies of scale, and is able to deal with multi-input, multi-output systems in which the factors have different scales. Using a data set of 43 large F/OS project retrieved from SourceForge.net, we have demonstrated the application of DEA. Results show that DEA indeed is usable for comparing the efficiency of projects in transforming inputs into outputs, that differences in efficiency exist, and that single projects can be judged and ranked accordingly. We have also detailed additional analyses based on these results, which stated that neither the inequality in work distribution within the projects, nor the licensing schemeor the intended audience has an effect on the efficiency of a project. As this is a first attempt at using this method for F/OS projects, several future research directions are possible. These include additional work on determining input and output factors and analyses based on the results using other project characteristics. These could include comparisons within application areas, different project scales, and comparisons to commercial or mixed-mode development projects.

References

Albrecht, A. J., & Gaffney, J. E. (1983). Software function, source lines of code, and development effort prediction: A software science validation. *IEEE Transactions on Software Engineering, 9*(6), 639-648.

Atkins, S., Ball, T., Graves, T., & Mockus, A. (1999). Using version control data to evaluate the impact of software tools. In *Proceedings of the 21st International Conference on Software Engineering*, Los Angeles, CA (pp. 324-333).

Banker, R. D., Charnes, A., & Cooper, W. (1984). Some models for estimating technical and scale inefficiencies in data envelopment analysis. *Management Science, 30*, 1078-1092.

Banker, R. D., & Kemerer, C. (1989). Scale economies in new software development. *IEEE Transactions on Software Engineering, 15*(10), 416-429.

Banker, R. D., & Slaughter, S. A. (1997). A field study of scale economies in software maintenance. *Management Science, 43*(12), 1709-1725.

Charnes, A., Cooper, W., & Rhodes, E. (1978a). *A data envelopment analysis approach to evaluation of the program follow through experiments in U.S. public wchool education* (Management Science Research Report No. 432). Pittsburgh, PA: Carnegie-Mellon University.

Charnes, A., Cooper, W., & Rhodes, E. (1978b). Measuring the efficiency of decision making units. *European Journal of Operational Research, 2*, 429-444.

Chidamber, S. R., & Kemerer, C. F. (1994). A metrics suite for object oriented design. *IEEE Transactions on Software Engineering, 20*(6), 476-493.

Cook, J. E., Votta, L. G., & Wolf, A. L. (1998). Cost-effective analysis of in-place software processes. *IEEE Transactions on Software Engineering, 24*(8), 650-663.

Cooper, W., Seiford, L., & Tone, K. (2000). *Data envelopment analysis: A comprehensive text with models, applications, references and DEA-solver software.* Boston: Kluwer Academic Publishers.

Crowston, K., Annabi, H., & Howison, J. (2003, December 14-17). Defining open source software project success. In *Proceedings of ICIS 2003*, Seattle, WA.

Crowston, K., Annabi, H., Howison, J., & Masango, C. (2004, May 25). Towards a portfolio of FLOSS project success measures. In *Collaboration, Conflict and Control: The 4th Workshop on Open Source Software Engineering, International Conference on Software Engineering (ICSE 2004)*, Edinburgh, Scotland.

Crowston, K., & Scozzi, B. (2002). Open source software projects as virtual organizations: Competency rallying for software development. *IEE Proceedings—Software Engineering, 149*(1), 3-17.

Dempsey, B. J., Weiss, D., Jones, P., & Greenberg, J. (2002). Who is an open source software developer? *Communications of the ACM, 45*(2), 67-72.

Dixon, R. (2003). *Open source software law*. Norwood, MA: Artech House.

Farell, M. J. (1957). The measurement of productive efficiency. *Journal of the Royal Statistical Society, Series A 120*(3), 250-290.

Ghosh, R., & Prakash, V. V. (2000). The Orbiten free software survey. *First Monday, 5*(7).

Hahsler, M., & Koch, S. (2005). Discussion of a large-scale open surce data collection methodology. In *Proceedings of the Hawaii International Conference on System Sciences (HICSS-38)*, Big Island, HI.

Hertel, G., Niedner, S., & Hermann, S. (2003). Motivation of software developers in open source projects: An Internet-based survey of contributors to the Linux kernel. *Research Policy, 32*(7), 1159-1177.

Holck, J., & Jorgensen, N. (2004). Do not check in on red: Control meets anarchy in two open source projects. In S. Koch (Ed.), *Free/open source software development* (pp. 1-26). Hershey, PA: Idea Group Publishing.

Hu, Q. (1997). Evaluating alternative software production functions. *IEEE Transactions on Software Engineering, 23*(6), 379-387.

Hunt, F., & Johnson, P. (2002). On the pareto distribution of Sourceforge projects. In *Proceedings of the Open Source Software Development Workshop*, Newcastle, UK (pp. 122-129).

Kitchenham, B. (2002). The question of scale economies in software—Why cannot researchers agree? *Information & Software Technology, 44*(1), 13-24.

Kitchenham, B., & Mendes, E. (2004). Software productivity measurement using multiple size measures. *IEEE Transactions on Software Engineering, 30*(12), 1023-1035.

Koch, S. (2004). Profiling an open source project ecology and its programmers. *Electronic Markets, 14*(2), 77-88.

Koch, S. (2005). *Effort modeling and programmer participation in open source software projects* (Arbeitspapiere zum tätigkeitsfeld informationsverarbeitung, informationswirtschaft und prozessmanagement, Nr. 03/2005). Department für Informationsverarbeitung und Prozessmanagement, Wirtschaftsuniversität Wien, Vienna, Austria.

Koch, S., & Schneider, G. (2002). Effort, cooperation and coordination in an open source software project: GNOME. *Information Systems Journal, 12*(1), 27-42.

Krishnamurthy, S. (2002). Cave or community? An empirical investigation of 100 mature open source projects. *First Monday, 7*(6).

Laurent, L. S. (2004). *Understanding open source and free software licensing.* Cambridge, MA: O'Reilly & Associates.

Mayrhauser, A., Wohlin, C., & Ohlsson, M. (2000). Assessing and understanding efficiency and success of software production. *Empirical Software Engineering, 5*(2), 125-154.

McCabe, T. J. (1976). A complexity measure. *IEEE Transactions on Software Engineering, 2*(4), 308-320.

Mockus, A., Fielding, R., & Herbsleb, J. (2002). Two case studies of open source software development: Apache and Mozilla. *ACM Transactions on Software Engineering and Methodology, 11*(3), 309-346.

Myrtveit, I., & Stensrud, E. (1999). Benchmarking COTS projects using data envelopment analysis. In *Proceedings of 6th International Software-Metrics-Symposium* Boca-Raton, FL (pp. 269-278).

Park, R. E. (1992). *Software size measurement: A framework for counting source statements* (Tech. Rep. No. CMU/SEI-92-TR-20). Software Engineering Institute, Carnegie Mellon University, Pittsburgh, PA.

Perens, B. (1999). The open source definition. In C. DiBona et al. (Eds.), *Open sources: Voices from the open source revolution.* Cambridge, MA: O'Reilly & Associates.

Raymond, E. S. (1999). *The cathedral and the bazaar.* Cambridge, MA: O'Reilly & Associates.

Robles, G., Koch, S., & Gonzalez-Barahona, J. M. (2004). Remote analysis and measurement of libre software systems by means of the CVSanalY tool. In *ICSE 2004—Proceedings of the Second International Workshop on Remote Analysis and Measurement of Software Systems*, Edinburgh, Scotland (pp. 51-55).

Rosen, L. (2004). *Open source licensing: Software freedom and intellectual property law.* Englewood Cliffs, NJ: Prentice Hall PTR.

Stallman, R. M. (2002). *Free software, free society: Selected essays of Richard M. Stallman.* Boston: GNU Press.

Stewart, K. J. (2004, May 25). OSS project success: From internal dynamics to external impact. In *Collaboration, Conflict and Control: The 4th Workshop on Open Source Software Engineering, International Conference on Software Engineering (ICSE 2004)*, Edinburgh, Scotland.

Stewart, K. J., & Ammeter, T. A. (2003). An exploratory study of factors affecting the popularity and vitality of open source projects. In *Proceedings of the 23rd International Conference on Information Systems*, Barcelona, Spain.

Varian, H. R. (2005). *Intermediate microeconomics: A modern approach* (7th ed.). New York: W. W. Norton & Company.

Weiss, D. (2005). Measuring success of open source projects using Web search engines. In *Proceeding of the 1st International Conference on Open Source Systems*, Genoa, Italy (pp. 93-99).

Section II

F/OSS Community Structures and Perception

Chapter III

Ecology and Dynamics of Open Source Communities

Michael Weiss, Carleton University, Canada

Gabriella Moroiu, Carleton University, Canada

Abstract

The goal of this chapter is to document the evolution of a portfolio of open source communities. These are communities formed around a set of related projects with common governance, which often produce artifacts shared among all projects. It helps to think of a portfolio of project communities as an ecology, in which the projects are mutually dependent, and there is both cross-project collaboration and competition for resources among the communities. As a case study, we explore the ecology of communities within the Apache project, one of the largest and most visible open source projects. We infer the community structure from developer mailing lists, and study how the communities evolve and interact over time. The analysis lends support to the often-stated hypothesis that open source communities grow by a process of preferential attachment. However, we show that the influx of external developers is not the only factor affecting community growth. The structure and dynamics of a community is also impacted by inter-project information flow, and the migration of developers between projects (including the formation of spin-offs).

Introduction

The open source model has received increasing attention as an alternative to closed source development. It is characterized by the transparency of development process and artifacts produced, as well as the decentralized organizational structure through which a community of developers and users coordinate their activities. The progress of an open source project is continuously tracked in a number of archives including code repositories, mailing lists, wikis, and bug tracking lists. Community members (both developers and users) can belong to any number of organizations, and their decision making process is governed by the principle of meritocracy, whereby members are given varying levels of access to the project based on their history of contribution.

Despite the evident successes of the open source model, however, we do not fully understand how open source communities organize themselves. We would like to gain a better understanding of the mechanisms underlying the growth of communities. From these, we could then derive guidelines for the design of successful open source communities. In this chapter we look at the evolution of community structure. As we want to understand the open source development model, we focus on developer communities. One goal of the chapter is to describe community structure in terms of observable metrics such as size or degree distribution. However, this does not provide us with insight on the inner workings of community organization. Therefore, we also explore potential growth factors and mechanisms that help explain the observed evolution of community structure.

As a case study, we explore the ecology of communities within the Apache project, both for reasons that this is a highly visible group of open source communities, but also because a wealth of data is being collected on the Apache project site that allows deep insight into the dynamic project structure. What sets this study apart from, for instance, a study of SourceForge projects (which largely vary in scope, size, and maturity) is that the Apache projects have a common domain (web services), are well established in the open source community, and are relatively large. For example, the observation made about SourceForge projects that most projects have only one developer and are not interconnected is not true about Apache projects. We have analyzed eight years (between 1997 and 2004) of project developer mailing lists to extract the structure of the associated communities. The selection of the Apache project, and the vehicle of mailing lists for recovering social interactions was also motivated by the availability of the Agora tool for visualizing community participation in the Apache project from project mailing lists, which was developed by an Apache core developer (Mazzocchi, 2006). However, as noted above, there are other ways of tracking community participation, and, ultimately, one would like to construct a model of community evolution by combining multiple views on community activity. Questions such as how contributions to code repositories and participation on mailing lists correlate await future research (e.g., Bird, Gourley, Devanbu, Gertz, & Swaminathan, 2006).

The chapter is structured as follows. First, we provide background on key concepts, and review related efforts to modeling community evolution. Then, we describe the methodology followed to extract community structure and the various indicators (such as developer rank) from the Apache mailing list archives. Subsequently, we show how the various communities within the Apache project evolve, and interact over time. We state our findings in the form of hypotheses, and present our evidence in their support. Finally, we discuss opportunities for future research, and offer concluding remarks.

Background

An *open source community* is a group of developers and users who share a common interest in a project, and who regularly interact with one another to share knowledge, and collaborate in the solution of common problems (Ye, Nakakoji, Yamamoto, & Kishida, 2005). Communities are at the core of what is described in (Gloor, 2006) as collaborative innovation networks (COINs), highly functional teams character-ized by the principles of meritocracy, consistency, and internal transparency. As shown in (Ye et al., 2005), an open source community co-evolves with its associ-ated project. A project without a community that sustains it is unlikely to survive long-term. As described in the next section, it will simply fail to attract developers. While our focus here is on developer participation, it is quite clear that without the participation of users, a community cannot be sustained either.

Members of an open source community play different roles, ranging from project leaders (maintainers) and core members (contributors) to active and passive users (van Wendel de Joode, 2003; Xu et al., 2004; Ye et al., 2005). Project leaders are often also the initiators of the project. They oversee the direction of the project, and make the major development decisions. Core members (contributors) are members who have made significant contributions to a project over time. Active users com-prise occasional developers and users who report bugs, but do not fix them. Passive users are all remaining users who just use the system. Core members can further be subdivided into creators (leaders) communicators (managers), and collaborators (Gloor, 2006).

Large open source projects such as Linux or Apache comprise many subprojects, not of all of which are strongly connected to one another. They are not associated with a single, homogenous community, but rather an *ecology* of communities (Healy & Schussman, 2003) is formed around these subprojects. However, the communities have a common governance (Apache Software Foundation for the Apache project),

Figure 1. Portfolio of communities in the Apache project and their relationships

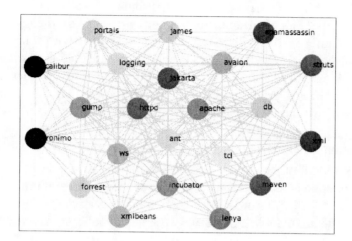

and often produce artifacts shared among all projects (such as the Jakarta Commons in Apache). The idea of ecology should convey mutual dependencies between many of the projects and cross-project collaboration, but also competition for resources among projects.

Figure 1 shows the current portfolio of communities in the Apache project and their relationships. It depicts the communication patterns between projects, as determined from the project mailing lists. In 2004, there were 23 projects, as identified by the mailing list names. This diagram was generated by an extension of the Agora tool (Mazzocchi, 2006), which reuses its data extraction and core visualization routines, but adds project and module dependency views, among others. The project communication view shows a link between two projects, whenever there are developers in both projects, who have exchanged messages on one of the associated developer mailing lists. Direct communication between projects cannot be measured, as mailing lists are project specific.

There is much anecdotal evidence that open source communities grow according to a preferential attachment mechanism, or *selection through professional attention* (van Wendel de Joode et al., 2003). However, there is not much empirical analysis to demonstrate this phenomenon. Most previous efforts have considered community evolution in a qualitative manner (van Wendel de Joode et al., 2003; Ye et al., 2005; Koch, 2005; Gonzalez-Barahona et al., 2004). One of the original motivations for our research, hence, was a discussion of how developers and users select open source communities in (van Wendel de Joode et al., 2003). The authors posit a selection mechanism that relies on two kinds of behaviors of community members: tagging and copying.

Tags are used as signals to attract other community members. An example of a tag is reputation. It is a sign to other developers that signals a certain level of knowledge or skill. Members of an open source community are inclined to copy the behavior of members with a high reputation. Thus, if a developer with a high reputation creates a project, other developers and users will be attracted to participate in this project. With a certain likelihood (also depending, of course, on other factors such as whether the specific project proves of interest to other community members) a community will form around the project. The size of a community then itself becomes a tag and signals the popularity of a project, in turn attracting further developers. As overall attention is limited, tagging and copying can explain the growth and decline of communities over time.

Our work is in a line of recent empirical studies of open source community evolution (Madey et al., 2005; Crowston & Howison, 2005; Roberts et al., 2006; Valverde & Solé, 2006). The main distinction of our work is that our approach is *constructive*, i.e., in addition to describing the observable behavior, we also aim to explore the underlying factors. A better understanding of those factors will allow us to guide the creation of successful new open source communities. As one example of the constructive approach, Robles et al. (2006) studied join/leave patterns in the evolution of project communities. They provide insight into the composition of development teams: Some projects are led by their original founders, while others have multiple generations of developers. Another example is provided by Sowe et al. (2006), who explore the role of knowledge brokers functioning as community facilitators, linking and collaborating with mailing list participants. Similar to our hypothesis regarding the migration of developers, Hahn et al. (2006) investigate the impact of social ties on open source community formation. They found that prior social ties of project initiators increase the likelihood of other developers joining a project. Our own previous results on the analysis of community evolution, presented in expanded form in this chapter, have been reported in Weiss et al. (2006).

As Robles et al. (2006) point out, the evolution of open source communities is a significant field to explore. In this chapter, *community evolution* refers to three aspects of community behavior: growth of communities, community life cycle, and migration between communities. As communities grow, they undergo a life cycle of formation, maintenance, and dissolution. In part, to gain an understanding of what is happening in a set of communities, we can look at the migration of its members between them. When members migrate, they may either leave the old community behind, or—by actively participating in both communities at the same time—maintain a bridge between the communities.

Much insight in how developers participate in an open source community can be gained by modeling them as *social networks*. In Madey et al. (2005) and Valverde & Solé (2006), a community is modeled as a network, in which nodes represent developers, and links or edges between nodes indicate that these developers participate in the same project. A characteristic of many of these networks is that they are dominated by a relatively small number of nodes linked to many other nodes. These nodes represent highly prolific developers or "hubs." Such networks are also known as scale-free networks (Barabasi & Bonabeau, 2003). Scale-free networks occur in many other kinds of complex systems, from the Internet, to authors citing each other's work, to movie actors.

Of particular interest is to understand why such networks give rise to hubs. A popular hypothesis to explain this phenomenon is that of *preferential attachment* (Barabasi et al., 2002). Intuitively, this hypothesis states that, as the network evolves, nodes will be more likely to link to nodes that already have a large number of links, or a high degree in the network. More generally, attachment can be preferential with regard to any property (such as degree or size) of the nodes attached to. For example, the more connected movie actors are more likely to be chosen for new roles. Such a "rich get richer" process tends to favor nodes that were added to the network early, as they are more likely to turn into hubs. The distribution of the number of nodes $P(k)$ with a given number of links k can often be observed to follow a *power law*: The probability that a node is connected to k other nodes is a continuously decreasing function. When plotted on a log-log scale, a power law will be a straight line, indicating that $P(k)$ is proportional to $k^{-\lambda}$ for some λ.

Methodology

The availability of open source project archives including code repositories, mailing lists, wikis, and bug-tracking lists presents an unprecedented opportunity for empirical research on community participation in open source communities (Mockus et al., 2005). Our study is based on extracting the structure of project communities from the associated mailing lists. As the Agora tool already collects the mailing lists for the Apache project, extracts the social interactions from them, and visualizes the structure of project communities, our emphasis is on the evolution of those communities.

Our work was guided by the following research questions:

1. How do open source communities evolve?
2. How do projects grow (what are the relevant growth factors)?
3. How do new projects form?
4. How does information flow between projects?

Figure 2. Extensions to the Agora community visualization tool

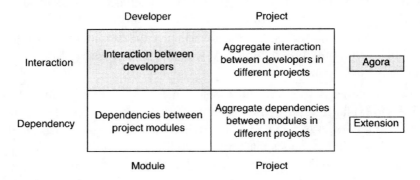

Research Approach

To trace the evolution of a community, we took snapshots of its membership at regular intervals. To this end, we extended the statistics capabilities of Agora and made its functionality scriptable, enabling the automated extraction of mailing list data. We also added project communication, and project/module dependency views to enable the study of aggregate behavior as well as the correlation between communication and code dependencies. Figure 2 describes the changes made to the Agora tool. In one dimension, we can study the interaction at the developer and project level; in the other, code dependencies at the module and project level. The latter is obtained using JDepend (2006). However, as we have not fully completed the extensions related to dependency evolution, this chapter will only use the communication views and the data we can extract from them.

We also built an exploratory tool in the rule-based language Prolog for rapidly modeling and testing new hypotheses about the extracted data. To this end, the data set is first converted into a set of Prolog facts. Hypotheses can be formulated iteratively by defining properties we wanted to infer as Prolog rules. For example, to study the migration of developers between projects, we define a set of rules that test whether, for a given pair of projects P and Q, a developer participates in project P in one year, and in project Q during the next one, but he or she is not already a member of project Q in the current year.

Data Extraction

The structure of a community can be inferred from the interactions between developers on the mailing list of the associate project. The algorithm used by Agora for

Figure 3. List of community members rank-ordered by frequency of contribution

Rank	Frequency				
r	f		rf = C	R(n)	
1 ceki	47	44	3	47	47
2 paul.smith	20	7	13	40	67
3 mwomack	17	7	10	51	84	1/3
4 rdecampo	11	10	1	44	95
5 hoju	10	13	-3	50	105
6 endre	8	5	3	48	113
7 rbair	6	5	1	42	119
8 robertburrelldonkin	5	5	0	40	124
9 oliver	4	5	-1	36	128
10 bloritsch	4	5	-1	40	132
11 rbair23	4	7	-3	44	136
12 nicolaken	4	4	0	48	140	2/3
13 kingjide_ii	4	2	2	52	144
14 siberski	3	3	0	42	147
15 log4j	3	4	-1	45	150
	206		68.6666667	137.333333	

extracting topological data from the message set is based on the concept of "reply": when a person sends a message in reply to another message, a link is created in the graph. Messages that are never replied to are considered noise, and excluded from the extracted data (Mazzocchi, 2006). From this information, a social network representing members can be constructed. For each developer, we tally the number of inbound and outbound messages. The project leader is considered the developer with the highest number of inbound messages, as this indicates how frequently this developer is consulted by others. It is, therefore, also a measure of the developer's reputation. The same metric is used in Gloor (2006) to identify creators, who provide the overall vision and guidance for a project.

Figure 3 shows a rank-ordered list of members of the logging project community (snapshot taken in 2003) sorted by frequency of their contribution. As can be seen, the majority of contributions to a mailing list are made by a small number of community members, collectively referred to as the core group. These comprise the previously identified roles of leaders and core members. In the example, one third of the contributions were made by only three developers, and the top 12 developers account for two thirds of the contributions.[1] The cut-off points at one third and two thirds are chosen based on Bradford's law, which in its original formulation describes the distribution of the top journals in a discipline, but has since been applied to many other fields. For details on Bradford's law and its application to open source communities, see Crowston, Wei, Li, and Howison (2006). Similar results have been obtained for other Apache projects. This confirms an observation made by Mockus et al. (2005) that the core group of a project has typically no more than 10-15 members.

Figure 4. Scripting interface to Agora

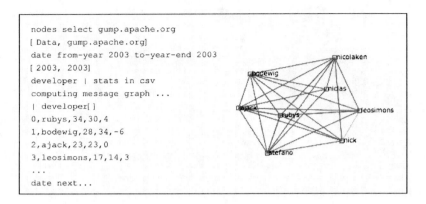

```
nodes select gump.apache.org
[ Data, gump.apache.org]
date from-year 2003 to-year-end 2003
[ 2003, 2003]
developer | stats in csv
computing message graph ...
 | developer[ ]
0,rubys,34,30,4
1,bodewig,28,34,-6
2,ajack,23,23,0
3,leosimons,17,14,3
...
date next...
```

Given that 10-15 members account for the majority of contributions to a community, we have limited most of our analyses to the top 15 members. This brings significant computational advantages during the analysis. Considering just the core group, we can derive insights about the evolution of communities that would generally be much more expensive to obtain, if all members were entered in the analysis. Of course, such an approximation requires validation. While we have not performed a formal analysis of the impact of this approximation, we have computed some of our results using both the core group and the full set of members, and have arrived at similar results.

The extended version of Agora can be used both interactively and through a scripting interface. While retaining the benefits of visual exploration of community structure, though scripting the extraction of data can be largely automated. For instance, the following sequence of commands selects the gump project (gump.apache.org), sets the time window to 1/2003-12/2003, and computes the developer message graph, as shown in Figure 4. Then, we ask Agora to produce a rank-ordered list of developers as described above. Such output can be redirected to a file for further processing. Finally, we move the time window to the next period of equal length, i.e.. 1/2004-12/2004.

A Perl script converts the generated data into a set of Prolog facts. An excerpt of the Prolog facts generated for snapshots of the gump project is shown in Figure 5. This representation corresponds to a model of community membership as a bipartite graph with nodes for projects and developers as used by Xu, Gao, Christley, & Madey, G (2005) and Sowe (2006). For further analysis, the extracted data can also be passed into a statistics tool like R.

Figure 5. Prolog facts generated from the extracted data

```
project(gump, 2002, []).
project(gump, 2003,['rubys','bodewig','ajack','leosimons','nicolaken','cmlenz',
'stefano', 'dion','nick','mllist','bloritsch','peter','vmassol','acoliver','dims']).
project(gump, 2004, ['ajack','bodewig','stefano','lsimons','niclas','antoine','nick',
'nicolaken','mcconnell','ceki','rubys','davanum','epugh','scott','michael.davey']).
```

Results

Growth by Preferential Attachment

Our initial hypothesis is that open source communities grow by a process of preferential attachment (Madey et al., 2005), or selection through professional attention (van Wendel de Joode, 2003). This hypothesis was adopted from the literature.

Hypothesis 1: *Larger communities will attract more new developers.*

We will approach this hypothesis in several steps. First, we want to show that open source communities can, indeed, be modeled as scale-free networks.

Hypothesis 1a: *Open source communities are dominated by a relatively small number of developers linked to many other developers.*

In support of hypothesis 1a, we first determine the distribution P(k) of the number of developers with degree k. This is the number of collaborations of a developer with other developers, as represented in the social network by links. As shown in Figure 6, this distribution follows a power law. This indicates that the communication network of the Apache community is scale-free. It has relatively few highly connected developers, while most developers are only connected to a few other developers. This leads to a typical core-periphery structure, as observed for many open source communities, with a group of core developers at the center, surrounded by less active developers and users.

As noted earlier, a common mechanism to explain the growth of a scale-free network is preferential attachment (Barabasi et al., 2002). Following the work of Pollner et al. (2006), we studied preferential attachment at two levels of granularity: at the developer level, and the project or community level. At this point, we cannot answer how precisely these attachment processes interact. An interesting question

Figure 6. Developer degree distribution P(k) shown with logarithmic binning

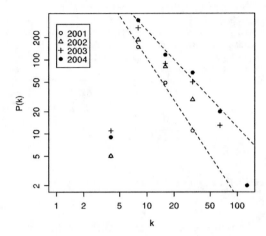

is to what extent the reputation of individual developers determines the selection made by new developers.

Hypothesis 1b: *New developers are more likely to link to well-connected developers. As a result, these well-connected developers become even more connected.*

To verify that the network of the Apache community follows a preferential attachment rule at the developer level, we first determine the probability $\Pi(k)$ that a new developer selects an existing developer with degree k. As described in Barabasi et al. (2002), this probability can be estimated by plotting the change in the number of links Δk for an existing developer over the course of one year as a function of k, the number of links at the beginning of each year. We expect $\Pi(k)$ to grow as k^v ($v>0$). Due to scatter, we plot its integral, the cumulative preferential attachment K(k), instead.[2] If attachment were uniform, K(k) would be linear ($v=0$). As shown in Figure 7, we find that K(k) is non-linear, and thus the probability $\Pi(k)$ of selecting existing developers increases with k. The exponent v is 0.38 with an adjusted R^2 of 0.893. This provides support for hypothesis 1b.

Having established that the growth of the Apache community follows a preferential attachment regime at the developer level, we repeat the analysis at the community level. At this level, we are concerned with links between communities. A link between two communities indicates that they share a developer. These relationships can be obtained from the project data by asking which developers participate in both projects.

Figure 7. Cumulative preferential attachment K(k) of new developers

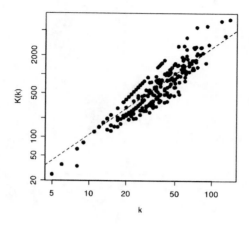

Hypothesis 1c: *New communities (which have not yet established links with other communities) are more likely to link to well-connected communities.*

Now, instead of estimating the probability of a new developer connecting to an existing developer, we determine the probability of a new community linking to an existing community. In order to show that this probability is proportional to the degree k^{com} of the existing community, we determine the change in the number of links for a community over the course of one year as a function of the number of links at the beginning of each year. Figure 8 shows the cumulative preferential attachment $K(k^{com})$ of a new community linking to an existing community. As shown, the exponent v is 0.75 with an adjusted R^2 of 0.925. Thus, as for the developer level, the probability increases with k^{com}. In support of hypothesis 1c, more connected communities attract more links from new communities.

This result obtains a deeper significance, if we note that community degree and community size are strongly correlated for higher degrees and larger sizes (Pollner et al., 2006). Therefore, an attachment process that is preferential with regard to community degree is also preferential with regard to community size. But this is what we had wanted to show in order to demonstrate hypothesis 1. The more developers a community has already, the more new developers it will attract. We took this indirect route, using community degree as a *proxy* for community size, because demonstrating hypothesis 1c for a given open source community implies hypothesis 1. However, hypothesis 1c is potentially much simpler to demonstrate, as we could obtain a reasonable estimate of community degree from just looking at the core group of a community. As noted, we have made this approximation to reduce the computational effort required for large networks. For the given network, we can also measure preferential attachment with regard to community size directly, using the

Figure 8. Cumulative preferential attachment K(k^{com}) of new communities

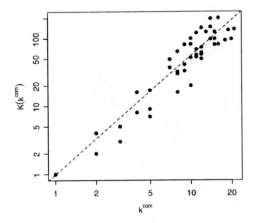

Figure 9. Cumulative preferential attachment K(s) as a function of community size

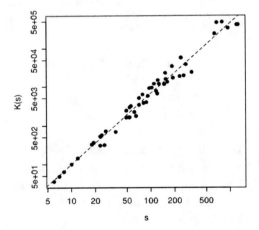

full data set obtained from the Agora tool, rather than cutting off the project size at 15. Figure 9 shows preferential attachment K(s) as a function of community size. The exponent v is 0.82 with an adjusted R^2 of 0.978, in close agreement with the value estimated from using community degree as a proxy for its size.

In summary, external growth of a community correlates with project degree and size. Yet, as much as the influx of external developers is a key characteristic of open source communities that distinguishes them from other types of networks, it is not the only factor that affects community evolution. As has been noted by Barabasi et al. (2002) and Pollner et al. (2006), the internal interaction between projects also affects the structure and dynamics of a community. Internal interaction comprises the flow of information, work products, and developers. We will look at the flow of information and developers next.

Information Flow

Information is shared between projects through common developers who act as bridges between the projects. In Gonzalez-Barahona et al. (2004), these developers are considered the "glue that maintains the whole project together, and the chains that contribute to spread information from one part of the project to another". The presence or absence of shared developers should have a visible impact on community growth.

Hypothesis 2: *Communities with more shared developers will interact more.*

In order to demonstrate this hypothesis, we measured the distribution of projects per developer, and inter-project preferential attachment. From what we know about scale-free systems, we expect to observe a small number of active developers that participate in many projects at the same time, while most developers participate in only a few.

Hypothesis 2a: *A small number of highly active developers participate in many projects at the same time, and facilitate inter-project information flow.*

Figure 10 shows the distribution of projects per developer during the year 2004. It follows a power law with a slope of -2.76 and an adjusted R^2 of 0.959. This result, similar to that obtained by others (e.g., Madey et al., 2005), means that while most

Figure 10. Distribution of the number of projects per developer (measured in 2004)

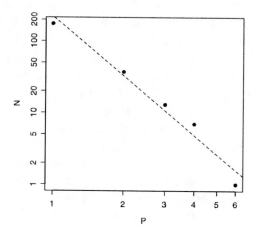

Figure 11. Cumulative internal preferential attachment $K(k_1{}^{com} k_2{}^{com})$ *between projects*

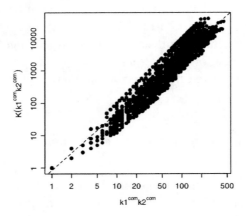

developers participate in only a few projects, a minority of them are active in many projects at the same time, as required by hypothesis 2a. These well-connected developers act as network hubs. Hubs connect many projects with one another, and are vital to the efficient functioning of the community. Yet, how do hubs select in which projects they should participate? An answer to this question can be obtained by looking for evidence of a preferential attachment process that governs the selection of projects by hubs.

For this purpose, we adopt an approach described by Barabasi et al. (2002) to model the creation of internal links in a scientific collaboration network, i.e., links between authors that are already part of the network, but have not collaborated before. We argue that the likelihood of creating links between two open source projects through shared developers depends on the links both projects have with other projects. One way of modeling the contributions of both projects, as does Barabasi et al. (2002), is to approximate the probability $\Pi(k_1{}^{com}, k_2{}^{com})$ that an existing project with degree $k_1{}^{com}$ and another project with degree $k_2{}^{com}$ form a new link as a function of the product $k_1{}^{com} k_2{}^{com}$. Indeed, when we plot the cumulative change $\Delta(k_1{}^{com} k_2{}^{com})$ as a function of $k_1{}^{com} k_2{}^{com}$, we observe a power law dependency as shown in Figure 11. This distribution measures the cumulative internal preferential attachment between projects, and has a slope $v=0.77$ and adjusted R^2 of 0.921 for the ecology of communities within the Apache project. It demonstrates that the number of shared developers grows according to a preferential attachment rule. Thus, the interaction of existing communities plays a critical role in community growth. While attracting new developers to an open source community is important for its continued survival, it is at least equally important to reduce the barriers for inter-community information flow. A simple way of estimating the interaction potential of two communities is to take the product of the number of developers they share with other communities.

Migration

As the previous sections demonstrated, preferential attachment governs the formation of both external and internal links in open source communities. Both mechanisms contribute to our understanding of the migration of developers between communities. Figure 12 summarizes what we have described so far in a model for migration in open source communities. This model introduces the concept of a "pool" that supplies new developers. Developers join a project either from the pool or from an existing project. When they move from one project to another, they can either leave the previous project, or become actively involved in both, thus facilitating the communication between the projects. Finally, we need to distinguish between newly formed and already existing projects.

What these mechanisms do not explain is how new communities are formed. We need to take a more detailed look at developer migration. To detect the migration behavior in the data, we look at pairs of projects, and test, for each pair P and Q, whether a developer participates in project P in one year and in project Q during the next one, but is not already a member of project Q in the current year. Figure 13 shows the developer migration from 2002 to 2003. Each row contains the number of developers migrating from a given project to any of the other projects during the following year. Note that "pool" is not a project, but indicates the influx of new core developers as in Figure 12.

Many of these developers migrate to new projects, of which they form the core to which new developers attach themselves. The columns in Figure 13 are listed in order of project creation. Thus, any project right of the column listing migrations into the maven project, that is, any project from gump to apachecon, was created in 2003. The numbers indicate the number of developers who have migrated. Developers can migrate to multiple projects, and each migration is given its own entry.[3]

Figure 12. Migration model for open source communities

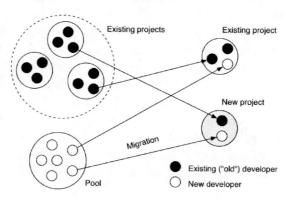

Figure 13. Migration between projects (transition from 2002 to 2003)

2002->2003	httpd	apache	jakarta	ant	apr	logging	ws	struts	tcl	avalon	incubator	xml	forrest	maven	gump	db	james	cocoon	portals	geronimo	xmlbeans	spamassassin	lenya	apachecon
	5	6	5	9	8	9	4	10	10	4	9	4	7		4	6	6	11	10	12	14	5	11	10
httpd																	1					1		2
apache						1									4	1	2	2				1	3	2
jakarta						1					3				4	2		1	1	1	1		1	1
ant						1						1			4			1	1			1		
apr																						1		1
logging		1									1													1
ws										1					3		4							
struts		1																1	1			1		
tcl																	2	1						
avalon		1				3									7		2	1				2		
incubator	2					1									3		2	1				1	1	
xml						1									3		3	1				3	2	
forrest						1									4		3						3	
maven		2				1					5				2	3			1	2	1			

As projects are spun off from existing projects, developers tend to migrate with community members they closely associate with. We should expect the effect to be most pronounced if the leader of one project moves on to a new project: This would create an even stronger pull for other core developers to join the new project. Thus, we surmise that developer *reputation* (e.g., as measured by developer degree) also plays a critical role in migration decisions.

Hypothesis 3: *Developers will migrate together with their collaborators.*

Figure 14 plots the distribution P(s) of group size s, that is, the number of developers migrating together to a new project from the *same* project. The distribution can be seen to observe a power law with a slope of -2.15 and R^2 of 0.982. The sample for this figure included all migrations between 1997 and 2004, excluding migrations from the pool. This result provides partial support for the hypothesis. It is indicative of the role of strong ties in determining migration behavior, but only accounts for the ties between developers within the projects from which they have migrated. A more extensive analysis will also need to take into account the prior history of collaboration among developers. Thus, while many developers will migrate in small groups, some well-connected developers will move in large groups, which provide the support for a new project. Our data support that most new projects include at least one large group migrated from another project. In passing, it may be observed that communities often start out small, and have even far less than 15 members. As noted in Gloor (2006), a typical core group starts out with three to seven members, and grows to 10 to 15 members once the community is established.

Future Trends

Current empirical research on the evolution of open source communities has taken two paths, often in conjunction with one another: measurement and model building, and simulation. Examples of measurement and model building include: Madey et al. (2005; Crowston & Howison (2005; Roberts et al. (2006; Valverde & Solé (2006; Robles et al. (2006; Gloor (2006). Examples of simulation include Madey et al. (2005); Valverde & Solé (2006). Simulation aims to replicate observed measurements. In her work on the co-evolution of epistemic networks, Roth (2005), although not explicitly talking about the evolution of open source communities, refers to this a "reconstructive" activity. Particularly promising seem agent-based simulation models. Madey et al. (2005) were among the first to apply agent models to reconstructing the behavior of open source communities.

Temporal analysis of interaction patterns in open source communities can also be used a a diagnostic tool to assess the health of an open source project. Gloor (2006) describes an environment for the visual identification and analysis of the communication in communities. Relating dynamic interaction patterns with the performance of work teams, his TeCFlow (temporal communication flow analysis) tool can identify typical communication patterns of communities and shed light on how well teams collaborate. We may also hope to derive general principles for managing the evolution of communities, that is, rules to be followed during the life cycle of successful open source projects.

Another fruitful venue for future research should be using the kind of quantitative analysis we have presented, in which we derive community properties from developer participation in mailing lists, as a way of narrowing down a qualitative analysis of interesting developer interactions. As presented, we are not making use of message subject and content, but some kind of interactions can only be studied by a more detailed manual analysis of the actual messages. Yet, such qualitative analysis is naturally limited in terms of the number of interactions to which it can be applied, and is therefore often very limited in scope. A hybrid approach, in which we first obtain a birds-eye view of community participation through quantitative analysis, and then follow up with a more detailed qualitative analysis for selected interactions, would resolve this tension.

Finally, developers of closed source development are paying close attention to the success of the open source model. Therefore, observations made about the evolution of open source communities should also be of great interest to organizations trying to apply these concepts to their internal development processes. Tools that measure and simulate the evolution of open source communities could potentially also be applied to the assessment and diagnosis of closed source development projects.

Conclusion

In this chapter we developed hypotheses about the evolution of a portfolio (or ecology) of related open source communities. To validate them we performed an empirical analysis of eight years of mailing list archives (between 1997 and 2004) for the communities within the Apache project. From these results we can conclude the following:

1. **How do open source communities evolve?** Open source communities evolve through a combination of two processes, at about equal rates,[4] by adding new developers from a pool of external developers (hypothesis 1), and through internal migration of developers between existing projects (hypotheses 2 and 3).

2. **How do projects grow (what are the relevant growth factors)? Growth** factors for open source communities include project degree and size: External growth correlates with project degree/size (hypotheses 1), and internal migration between two projects follows the product of the project degrees (hypothesis 2).

3. **How do new projects form?** Project degree/size cannot explain the formation of new projects. Thus, we also studied migration patterns of developers, leading to the insight that new projects form around a core of well-connected developers (hypothesis 3). This may help explain how such projects manage to grow.

4. **How does information flow between projects?** Shared developers facilitate the information flow between projects (hypotheses 2 and 3). As hypothesized by Weiss et al. (2006), information flow may also follow dependencies in the project code, that is, the communication and code structure of projects should be reciprocal.

In future work, we hope to complement our statistical analysis through simulation. The simulation will allow us to verify and validate our models of the evolution of open source communities, and help identify new factors to be included in those models. We will also work towards distilling the observations into actions for the management of open source projects. In conjunction with this, we want to explore the potential of using our models as diagnostic tools for assessing the health of an open source project.

References

Barabasi, A., Jeong, H., Neda, Z., Ravasz, E., Schubert, A., & Vicsek, T. (2002). Evolution of the social network of scientific collaborations. *Physica A 311*, 590-614.

Barabasi, A. L., & Bonabeau, E. (2003, May). Scale-free networks. *Scientific American*, 60-69.

Bird, C., Gourley, A., Devanbu, P., Gertz, M., & Swaminathan, A. (2006). Mining email social networks. *International Workshop on Mining Software Repositories* (pp. 137-143).

Crowston, K. & Howison, J. (2005). The social structure of open source software development teams. *First Monday, 10*(2).

Crowston, K., Wei, K., Li, Q., & Howison, J. (2006). Core and periphery in free/ libre and open source software team communications. In *Proceedings of the International Conference on System Sciences* (pp. 118-124). IEEE.

Feller, J., Fitzgerald, B., Hissam, S., & Lakhani, K. (2005). *Perspectives on free and open source software*. MIT Press.

Gloor, P. (2006). *Swarm creativity*. Oxford University Press.

Gonzalez-Barahona, J., Lopez, L., & Robles, G. (2004). *Community structure of modules in the Apache project*. Presented at the Workshop on Open Source Software Engineering.

Hahn, J., Moon, J. Y., & Zhang, C., Impact of social ties on open source project team formation. In E. Damiani, B. Fitzgerald, W. Scacchi, M. Scotto, G. Succi (Eds.), *Open source systems*, Como Italy (Vol. 203, pp. 307-317). Springer.

Healy, K., & Schussman, A. (2003). *The ecology of open source development*. Unpublished. Retrieved from www. kieranhealy.org/files/drafts/oss-activity.pdf

JDepend. (2006). *Project*. Retrieved July 2006, from www.clarkware.com/software/JDepend.html

Koch, S. (2005a). *Free/open source software development*. Hershey, PA: Idea Group Publishing.

Koch, S. (2005b). Evolution of open source software systems—A large-scale investigation. *International Conference on Open Source Systems* (pp. 148-153).

Madey, G., Freeh, V., & Tynan, R. (2005). Modeling the F/OSS community: A quantitative investigation. In S. Koch (Ed.), Free/open source software development (pp. 203-221). Hershey, PA: Idea Group Publishing.

Mazzocchi, S. (2006). *Apache Agora 1.2. Tool.* Retrieved July, 2006, from people. apache.org/~stefano/agora/

Mockus, A., Fielding, R., & Hersleb, J. (2005). Two case studies of open source software development: Apache and Mozilla. In Feller et al. (pp. 163-209).

Pollner, P., Palla, G., & Viczek, T. (2006). Preferential attachment of communities: The same principle, but at a higher level. *Europhysics Letters, 73*(3), 478-484.

Roberts, J., Hann, I.-H., & Slaughter, S. (2006, June 8-10). Communication networks in an open source project. In E. Damiani, B. Fitzgerald, W. Scacchi, M. Scotto, G. Succi (Eds.), *Open source systems*, Como Italy (Vol. 203, pp. 297-306). Springer.

Robles, G., & Gonzalez-Barahona, J. (2006). Contributor turnover in libre software projects. In E. Damiani, B. Fitzgerald, W. Scacchi, M. Scotto, G. Succi (Eds.), *Open source systems*, Como, Italy (Vol. 203, pp. 273-286). Springer.

Roth, C. (2005). *Co-evolution in epistemic networks: Reconstructing social complex systems*. PhD Thesis, Ecole Polytechnique, Paris, France.

Sowe, S., Stamelos, I., & Angelis, L. (2006). Identifying knowledge brokers that yield software engineering knowledge in OSS projects. *Information and Software Technology, 48*, 1025-1033.

Valverde, S., & Solé, R. (2006). Evolving social weighted networks: Nonlocal dynamics of open source communities. *Europhysics Letters, 60*.

van Wendel de Joode, R., de Bruijn, J., & van Eeten, M. (2003). Protecting the virtual commons. In T. M. C. Asser (Ed.), *Information technology & law series*, 44-50.

Weiss, M., Moroiu, G., & Zhao, P. (2006). Evolution of open source communities. In E. Damiani, B. Fitzgerald, W. Scacchi, M. Scotto, G. Succi (Eds.), *Open source systems*, Como Italy (Vol. 203, pp. 21-32. Springer.

Xu, J., & Madey, G. (2004). *Exploration of the open source software community*. Paper presented at NAACOSOS Conference.

Xu, J., Gao, Y., Christley, S., & Madey, G. (2005). A topological analysis of the open source software development community. In *Proceedings of the 38th Annual Hawaii International Conference on System Sciences*, (Track 7, Vol. 7, pp. 1-10). Washington, DC: IEEE Computer Society.

Ye, Y., Nakakoji, K., Yamamoto, Y., & Kishida, K. (2005). The co-evolution of systems and communities in free and open source software development. In S. Koch (Ed.), Free/open source software development (pp. 59-82). Hershey, PA: Idea Group Publishing.

Endnotes

[1] In the figure, R(n) is the cumulative number of messages received. The total number of messages is 206, and 68 and 137 are the respective ideal cut-off values for R(n).

[2] Note that if $\Pi(k)$ grows as k^v, $K(k)$ will grow as k^{v+1}, and to estimate v, we will need to subtract 1 from the exponent obtained from the regression for $K(k)$.

[3] An interesting topic for future work is to study the assignment of different weights to migrations, which could be derived from the rank the developer has in each project.

[4] This may only be true of the Apache project. However, as Robles et al. (2006) have documented, developer turnover is a significant feature of open source communities, and thus the behavior of open source communities may differ from that observed in other types of social networks, such as scientific collaboration networks, for which Barabasi et al. (2002) have noted that growth is governed largely by internal interactions.

Chapter IV

Perceptions of F/OSS Community:
Participants' Views on Participation

Andrew Schofield, University of Salford, UK

Grahame S. Cooper, University of Salford, UK

Abstract

The role of online communities is a key element in free and open source software (F/OSS) and a primary factor in the success of the F/OSS development model. F/OSS communities are inter-networked groups of people who are united by a common interest in F/OSS software. This chapter addresses holistic issues pertaining to member participation in F/OSS communities, specifically considering their reasons and motivation for participating. It collates the relevant literature on F/OSS community participation and presents the results of an empirical study into members' perceptions of their own participation. We identify primary reasons for participation such as problem solving, support provision, and social interaction and rank their importance by the participants' preferences. We then separate development and support activities and compare the community members' perceptions of the two. Finally, we draw conclusions and discuss the potential for future research in this area.

Introduction

The vast expansion of the Internet and the proliferation of electronic communication techniques have brought about many changes in the way people live and work. It has allowed people from around the world to communicate freely and easily by breaking down many of the geographical barriers that once stood in their way. These technological developments have also led to the transposition of social activities into a new electronic form. Several authors (Gattiker, 2001; Rheingold, 2000; Ribeiro, 1997; Schoberth, Preece, & Heinzl, 2003; Smith & Kollock, 1994; Wellman & Gulia, 1995) have written on the subject of the online or virtual community. A huge subject area in itself, this literature examines how social behaviour adapts when the Internet is used as a communication medium. Free and open source software (F/OSS) is also intrinsically linked with the development of the Internet. The relationship is symbiotic, as much of the underlying software that makes up the Internet is F/OSS, and yet F/OSS relies on the Internet for the dissemination of software, and communication between developers and users. This is where these two concepts of virtual community and F/OSS meet. However, F/OSS communities differ from other types of virtual communities because of their emphasis on software. Likewise F/OSS development differs from traditional software development, largely due to the use of the Internet as a development forum. F/OSS community is therefore a unique phenomenon, the details of which can appear undefined and illusive.

An investigation into the social aspects of F/OSS communities must take into account the views of their participants. This chapter attempts to do just that by first analysing the literature on the subject of F/OSS communities, specifically addressing issues of motivation in detail, and then presenting the results of survey research on community participants' perceptions of their own participation.

Background

There has been a significant amount of research conducted which has attempted to properly define what is meant by the term *online community*. Rheingold's (2000) work on what he called "the virtual community" presents a definition stating that these communities are "social aggregations" which are created only when there is sufficient social interaction. The emphasis in this definition is clearly on the social theme. Along with this definition, Rheingold provides a large list of the reasons why people might become part of a virtual community, which is too lengthy to describe here in detail. However, reasons range from finding friends to conducting commerce and from playing games to chatting. Since Rheingold's work, however, the types of virtual community, or online community as they are often referred to, have become more distinct. The majority of Internet users are now familiar with commerce communities such a www.ebay.com and www.amazon.com, and the many Web log community sites. F/OSS communities are also now widely acknowledged, as are other types of communities, such as those based around online gaming. The latter is an interesting example, especially the communities based around massively multiplayer online role-playing games (MMORPGs) (Bartle, 2003; Ducheneaut, Yee, Nickell, & Moore, 2006; Yee, 2006). Although F/OSS and gaming communities exist for different reasons, certain parallels can be seen when examining literature from the two research areas. They both consist of Web-based forums where discussions between members take place and they both revolve around the core activity of software development and gaming respectively. When one examines this in more detail, however, there are some aspects that clearly highlight the differences between the two types of community. O'Mahony and Ferraro (2004, p. 10) state the following about F/OSS communities: "Unlike internet communities focused on hobbies, fantasy, gaming or social support, these communities are more like distributed project teams in a production environment."

Although this statement is true of the development-related activities of F/OSS communities, it is likely that the social sides of the two communities are very similar. Furthermore, it seems fair to postulate that the social aspects of most online communities will have common trends (Hildreth, Kimble, & Wright, 1998; McKenna & Bargh, 2000; Schoberth et al., 2003; Walther, 1996). There is specific literature within the online gaming domain which specifically addresses issues of motivation, such as Bartle (2003) and Yee (2006). However the motivations suggested mainly apply to the gaming itself. All of this suggests that although a significant amount is now known about individual community types, further research should be conducted to examine correlations between them.

As is true for all online communities, for a F/OSS community to exist there must be some form of common media to enable communication between members (Schoberth et al., 2003). While some authors (Lanzara & Morner, 2003; Oh & Jeon, 2004; Raymond, 2000) describe F/OSS communities as entirely virtual systems that operate almost exclusively over the Internet on a global scale, others (Krishnamurthy, 2002; O'Mahony & Ferraro, 2004) maintain that in the case of many communities, much of the communication and collaboration takes place off-line in the "real world" and that a significant amount of F/OSS development is not performed by groups but by individuals. In actuality, it is likely that both these views are true in some cases and that communities vary significantly from one to the next. While some F/OSS projects will have many people involved, others will have a small number or a single developer. Furthermore, some communities will communicate and collaborate exclusively online, while others will have offline meetings or even exist entirely offline. It is highly likely that offline meetings will be more common in projects taking place within organisations and between core members of a project (Schofield & Mitra, 2005).

Placing F/OSS projects and communities into categories can be difficult, due to the many variables that need to be taken into consideration. Likewise, the community members themselves cannot easily be categorised due to the nature of the online community. Research by Zhang and Storck (2001) illustrates this issue by creating the definition of "peripheral members," that is, members who may not participate directly with the community. If members do not participate by posting on discussion boards or mailing lists, by e-mailing other members, or by making other named contributions, then they are effectively invisible to other members. The term *lurker* has become a common description of this kind of user. They may visit forums and read what others have said, or receive the e-mails from a mailing list, but this does not require any real involvement or interaction. Clearly this raises questions about the definition of a community member. Is a person who visits a community forum to read what others have said actually a member of that community? Are the number of visits a factor? Is some interaction needed and, if so, how much? These are all interesting questions that researchers have attempted to answer but for now the only thing that is known for certain is that many people do have some involvement with F/OSS communities but do not interactively participate. It is interesting to consider this in comparison to the work of other authors (O'Mahony, 2004; Sagers, 2004) who believe that the social interaction that takes place within F/OSS communities is the core activity and the very foundation of their existence. This suggests that despite a large number of non-interactive members, a critical mass of participating members is needed for the community to exist and function.

How a member may communicate and collaborate with their community is dependant on the available interaction mechanisms that are in place and the specific needs of the particular member. These needs are very important factors and there are myriad reasons why a person will get involved with a F/OSS community. These issues of motivation have inspired several authors to research this area. The majority of the research conducted, however, has focused on the motivation of F/OSS developers in an attempt to clarify why they participate in F/OSS development projects. Although it could be considered the primary activity, software development is not the only thing happening within F/OSS communities. Significantly less research has attempted to take a more holistic view of F/OSS communities, that is, not just the development aspects. Research which has been carried out on the non-development activities mostly focuses on what is generally considered to be the secondary activity of F/OSS communities, namely, usage support provision and acquisition. It is interesting that at least three papers that the authors are aware of (Lakhani & von Hippel, 2003; Sowe & Stamelos, 2005; Sowe, Stamelos, & Angelis, 2006), refer to these activities as "mundane." Nevertheless, these support activities are clearly essential for F/OSS to survive, especially as other support mechanisms, such as those provided with most proprietary software, are often not available with F/OSS.

Table 1. Taxonomy of motivations in literature

Category	Subcategory	Motivation	Paper Ref.
		Development	
Technological	Self-efficacy	To meet personal technological need	Feller and Fitzgerald (2002)
		"Scratching an tch"	Raymond (2000)
		Personal needs	Hars and Ou (2001)
		Personal programming needs	Edwards (2001)
		Improve software for one's own use	Hertel et al. (2003)
		Own use of improved software	Lerner and Tirole (2002)
		Code needed for user need	Lakhani and Wolf (2003)
		Direct need for software and/or improvements	van Wendel de Joode et al. (2003)
		Use value/need for OSS product functionality	Hann et al. (2004)
		To exploit the efficiency of peer review	Feller and Fitzgerald (2002)
		Get help in realizing a good idea for a software product	Ghosh et al. (2002)
	Pragmatism	Facilitating daily work with software	Hertel et al. (2003)
		To work with "bleeding edge" technology	Feller and Fitzgerald (2002)
		Control over technology	von Krogh et al. (2003)
		To exchange knowledge	Ghosh et al. (2002)
		Solve a problem that could not be solved by proprietary software	Ghosh et al. (2002)
		Problem solving	Moon and Sproull (2000)
Economic	Corporate/ work place related	Benefits for developer's firm	Lerner and Tirole (2002)
		Securing venture capital	Lerner and Tirole (2002)
		Work need	Lakhani and Wolf (2003)
	Monetary	Make money	Ghosh et al. (2002)
		To strike it rich through stock options	Feller and Fitzgerald (2002)
		Revenues from related products and services	Hars and Ou (2001)
		Improved payment	Hertel et al. (2003)
		Alumni effect/reduced cost	Lerner and Tirole (2002)
	Economic pragmatism	Costs are low and benefits are high	van Wendel de Joode et al. (2003)
		Low opportunity cost, nothing to lose	Feller and Fitzgerald (2002)
		"Might as well make it OS"	Hars and Ou (2001)
		Distribute unmarketable software products	Ghosh et al. (2002)

continued on the following page

Table 1. continued

		Career concerns	Hann et al. (2004)
Economic	Career orientated	To gain future career benefits	Feller and Fitzgerald (2002)
		Self-marketing	Hars and Ou (2001)
		Career advantages	Hertel et al. (2003)
		Career incentives	Lerner and Tirole (2002)
		Improve my job opportunities	Ghosh et al. (2002)
		Enhance professional status	Lakhani and Wolf (2003)
		Accrued reputation	Moon and Sproull (2000)
	Skill acquisition & development	Improve programming skills	Lakhani and Wolf (2003)
		To improve coding skills	Feller and Fitzgerald (2002)
		Human capital	Hars and Ou (2001)
		Improving one's own programming skills	Hertel et al. (2003)
		Learn and develop new skills	Ghosh et al. (2002)
		Learning opportunities	von Krogh et al. (2003)
		To learn and share what is known	Scacchi et al. (2006)
		Share knowledge and skills	Ghosh et al. (2002)
Socio-political	Reputation	Gaining reputation	Hertel et al. (2003)
		Reputation	Kollock (1999)
		Reputational benefits	Lerner & Tirole 2002
		Reputation among hackers	Raymond 2000
		Get a reputation is OS/FS Community	Ghosh et al 2002
		Enhance reputation in F/OSS community	Lakhani & Wolf 2003
		Recognition of work	Moon & Sproull 2000
		Recognition as trustworthy and reputable contributors	Scacchi et al 2006
		Reputation	von Krogh et al 2003
		Reputation/garner stature within OSS Community	Hann et al 2004
		Influence and reputation/intrinsic reward logic	Edwards 2001
	Identification	Following other members/ membership herding	Oh & Jeon 2004
		Community Identification	Hars & Ou 2001
		General Identification as Linux developer	Hertel et al 2003
		Like working with this development team	Lakhani & Wolf 2003

continued on the following page

Table 1. continued

Socio-political	Commitment /reciprocity	Rewards from collective action	von Krogh et al 2003
		Expecting Reciprocity	Moon & Sproull 2000
		Reciprocity	Kollock 1999
		Feel personal obligation to contribute because use F/OSS	Lakhani & Wolf 2003
		Commitment & reciprocity	O'Mahony & Ferraro 2004
	Social reward	Sense of belonging to a community	Feller & Fitzgerald 2002
		Attachment to community	Kollock 1999
		Peer Recognition	Hars & Ou 2001
		Ego-satisfying piece of the action	Raymond 2000
	Self gratification	Ego gratification and signalling incentives	Feller & Fitzgerald 2002
		Satisfaction and fulfilment	Hars & Ou 2001
		Self-efficacy	Kollock 1999
		Personal satisfaction	Gacek & Arief 2004
	Hedonistic	Hobby	Edwards 2001
		Enjoying the work	Scacchi et al 2006
		Intrinsic motivation of coding	Feller & Fitzgerald 2002
		Fun to program	Hars & Ou 2001
		Enjoyment of the work itself	van Wendel de Joode et al 2003
		Having fun programming	Torvalds & Diamond 2001
		Intrinsic pleasure/for the joy of it	Moon & Sproull 2000
		Code for project is intellectually stimulating to write	Lakhani & Wolf 2003
	Leadership	Being own Boss	Lerner & Tirole 2002
		Leading a community/project	Lerner & Tirole 2002
	Altruism	Altruism	Feller & Fitzgerald 2002
		Altruistic values	Hars & Ou 2001
		Improve OS/FS products for other developers	Ghosh et al 2002

continued on the following page

Table 1. continued

Socio-political	Adversarial	Competitiveness with other developers and/or projects.	Bezroukov 1999
		Limit the power of large software companies	Ghosh et al 2002
		Dislike proprietary software and want to defeat them	Lakhani & Wolf 2003
	Idealism	OS Idealism	Hars & Ou 2001
		Software should be free	Hertel et al 2003
		Think that software should not be a proprietary good	Ghosh et al 2002
		Belief in free software	Elliot 2005
		Believe that source code should be open	Lakhani & Wolf 2003
		Software's Openness	Gacek & Arief 2004
		Being part of a social movement	von Krogh et al 2003
		Participate in a new form of cooperation/in the OS/FS scene	Ghosh et al 2002
	Social	Social networking	Hars & Ou 2001
		Personal exchange	Hertel et al 2003
		To take part in the main communications and discussion	Ghosh et al 2002

Non-Developer /Support			
Technological	Pragmatism	Primary means to get product feedback and updates	Sowe et al. (2006)
		To use code	van Wendel de Joode et al. (2003)
		To be notified of things happening in the community	van Wendel de Joode et al. (2003)
		Easy access to software	van Wendel de Joode et al. (2003)
	Support provision	Provide support because I have experience in this area	Lakhani and von Hippel (2003)
		Provide support because I am the authority in this area	Lakhani and von Hippel, (2003)
		Answer questions relating to packages I maintain	Sowe et al. (2006)
		Answer questions relating to other packages	Sowe et al. (2006)
		Answer questions relating to my specialized area	Sowe et al. (2006)
		To provide the best answer to a question	Lakhani and von Hippel (2003)

continued on the following page

Table 1. continued

Technological	Obtaining support	Getting help with critical problems	Lakhani and von Hippel 2003)
		Reporting a problem with favourite software	Edwards (2001)
		Reporting a bug or problem	van Wendel de Joode et al. (2003)
		Participate in development discussions	von Krogh et al. (2003)
Economic	Career orientated	I answer to enhance my career prospects	Lakhani and von Hippel (2003)
		I answer because it's my job	Lakhani and von Hippel (2003)
		Learning opportunities from lurking	von Krogh et al. 2003
		Free download of software	van Wendel de Joode et al. (2003)
		To learn about OS for corporate needs	van Wendel de Joode et al. (2003)
Socio-political	Identification/ Reputation	Community identification	Hertel et al. (2003)
		To gain reputation	Lakhani and von Hippel, (2003)
		General identification as Linux user	Hertel et al. (2003)
	Social reward	Copying the behaviour of others with good reputation	van Wendel de Joode et al. (2003)
		Involvement in meritocracy	O'Mahony & Ferraro (2004)
	Idealism	Supporting software and community	Hertel et al. (2003)
		Promotion of OSS	Lakhani and von Hippel, (2003)
	Support reciprocity	Expecting reciprocity (I help so I will be helped)	Lakhani and von Hippel, (2003)
		I was helped so I help	Lakhani and von Hippel, (2003)
	Hedonism	I answer because it's fun	Lakhani and von Hippel, (2003)
		Intrinsic motivation of answering questions	Lakhani and von Hippel, (2003)

continued on the following page

To summarise the previous work in this area, Table 1 presents a taxonomy of participant motivations identified in the literature, for both development and non-development/support activities, along with references to the source work. To facilitate the creation of this taxonomy, the framework developed in Feller and Fitzgerald (2000), and applied to the motivational drivers of F/OSS developers in Feller and Fitzgerald (2002), has been used as a basis. The framework splits motivation into three categories: technological, economical, and socio-political. Using these categories, it then applies them to the micro level, which addresses the individual, and the macro level, which addresses the organisation or community. Due to the aims of this chapter and the amount of information available, only the micro level has been targeted in the present taxonomy. The categories have been further broken down into sub-categories based on the motivations found in the literature. Note that the same three major categories defined by Feller and Fitzgerald have been applied to the literature pertaining to non-developer activities.

It is acknowledged that these categories may not be an exhaustive list and that many of the motivations listed may overlap and/or apply to other categories than the ones they are listed in. Nevertheless, the taxonomy attempts to identify the primary motivations of F/OSS developers and non-developer participants from the literature. Compiling the list was not easy, as the research in this area has been conducted using a variety of methods, from sociological to economic. We have tried to include only literature that explicitly states motives and that uses proven research results. We have also attempted to keep the listed motivations in their original form with as little interpretation as possible.

Although the literature listed in the taxonomy does provide valuable data and conclusions on the motivation of F/OSS community participants, in general it does not address the more holistic issues of the motivation of F/OSS participants. This chapter attempts to contribute to the filling of this knowledge gap by presenting the results of a research project which aimed to define community members' perceptions of their environments and their own participations.

Research Method

Online communities are intrinsically difficult to research, primarily due to their constant changing membership. In most cases a researcher cannot know exactly who is in the community, or how many subjects are under examination, and therefore cannot be sure how representative their sample is. This problem is exacerbated by the use of any research method that does not involve personal interaction with the sample. This research project used an online survey technique to collect data on F/OSS communities. The survey was designed in such a way that it would return predominantly quantitative data but also some supporting qualitative data. To increase the likelihood of people completing the survey, most questions were in multiple-choice format, with additional space provided for respondents to input their own answer or to write additional comments. The latter was provided in an attempt to prevent the researcher from leading the respondent and to allow for the collection of qualitative data.

The sample set used for the research was open source user groups, consisting for the most part of Linux and BSD user groups. Hereafter we shall refer to these groups as Linux User Groups or LUGs. The majority of a LUG's members will almost certainly participate with other F/OSS communities and therefore a member of a LUG is not exclusively involved with Linux or BSD. The results of the survey have shown that many members are involved with other independent F/OSS projects. It was these characteristics that the survey exploited, as it was not aimed at collecting data about the members' perceptions of the LUGs themselves but rather of F/OSS communities in general. Analysis of the responses showed that 12% of recipients were not even part of a F/OSS/Linux user group, club, or society. As the request for participation was only sent to LUG members, this statistic demonstrates the inter-connective nature of F/OSS communities. Initially, the survey was directed at LUGs based in the United Kingdom, which resulted in 145 submissions being received. The dissemination was then expanded to include the countries with the most LUGs, that is, the United States of America, Germany, Italy, and Canada. A further 289 responses were consequently received, making a total of 434 submissions. The survey was designed to collect members' perceptions of F/OSS communities. Specifically, the questions attempted to collect data on the activities members are involved with and how, and for what purpose, they interact and make use of communities.

Research Findings

Anyone who participates in a F/OSS community in any way will have some reason for doing so. This may be to make use of a function that the community provides or to carry out some activity supported by the community. It could be the use of an online tool or interaction with a particular discussion taking place within a F/OSS community forum. The first section of the survey dealt with these issues by collecting members' perceptions of what they do within F/OSS communities and their reasons for doing so. It is interesting here to note the difference between a participant's motivation to take part in a F/OSS community and their reasons for doing so. The distinction between the motivating factor and the direct pragmatic reason for participating is somewhat blurred. The majority of the literature on motivation, particularly the work of Hann, Roberts, and Slaughter (2004); Hertel, Niedner, and Herrmann(2003); Lakhani and Wolf(2003), and Scacchi, Feller, Fitzgerald, Hissam, and Lakhani (2006), concentrates on the initial motivation. For example, "receiving recognition from peers" could be the initial motivation for a member to participate in the apparently pragmatic activities of problem solving and providing support. A developer may participate within a community for pragmatic reasons of software development but their more general motivation may be to learn and develop their skills for their own career advancement. The important point is that the dividing line cannot always easily be drawn between purpose and motivation. This research has attempted to focus more on the members' direct reasons for participating and what they actually do within F/OSS communities.

As with all the survey questions, respondents were given the option to provide additional information or alternative answers to the multiple choices provided for them. For the first phase of the survey, participants were given the following options as to why they make use of F/OSS communities. It was possible for them to choose more than one option from the following:

- To find out how to perform a task in a software application (problem solving)
- To help other people to use software applications (providing support)
- To suggest alterations or improvements to software programs (peer review)
- To contribute bug fixes or code improvements (software development)
- To meet people or talk to people with similar interests (social exchange)

Figure 1 shows the responses as the actual number of choices. Out of the 434 participants, 372 of them (85.7%) chose problem solving as one of their reasons for participating. This was closely followed by providing support, which 343 participants

Figure 1. Reasons for participating

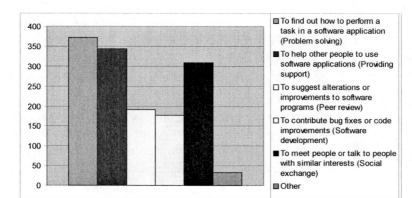

(79%) chose. The third most popular choice was social exchange, selected by 309 participants (71.2%). Interestingly, a total of 306 participants (70.5%) chose both problem solving and providing support, and a total of 231 (53.2%) chose problem solving, providing support, and social exchange. Peer review and software development were significantly less popular choices. However, this was expected, as not all F/OSS community participants are developers. Nevertheless, 192 participants (44.2%) selected peer review and 178 (41.0%) selected software development as a reason for participating.

Qualitative data collected from the respondents was also analysed and showed other reasons for participating. Many of these responses were duplicated, indicating that there are common issues among community members. Qualitative responses, presented here as generalised categories, were:

- **Community leadership:** Many of the respondents were responsible for a F/OSS community and therefore their role was one of management. One specific point was that development project leaders often need support from other communities developing related software (e.g., libraries). Other comments were made about the use of the communities as a tool for making announcements. This is especially relevant for the LUGs as they often organise meetings and other events.

- **Project recruitment/promotion:** Some respondents used F/OSS communities to get in touch with potential developers and/or users. This is effectively the advertising of a project or piece of software through a community.

- **Social/business networking:** Many survey participants wrote of building business relationships through F/OSS communities. More, however, emphasised the social networking that occurs through meeting people with similar interests. Writing of the social aspects, one participant described F/OSS communities as "just another kind of pub."

- **General interest and learning:** As a distinct goal from problem solving, learning about F/OSS was the reported goal of many participants. It was the view of many that F/OSS communities are the ideal place to learn about software development and how things are done within the F/OSS domain. Many participants also wrote of how they used F/OSS communities to stay up to date with the latest F/OSS news and developments which they would otherwise not be aware of.

- **Education:** Apart from members using the community to learn, some use them to teach. Some respondents wrote that they host courses within F/OSS communities. This results in a more formal learning environment than being immersed in a F/OSS community and picking things up as one goes along.

- **Advocacy:** Demonstrating the strong belief in the ideals of F/OSS, many respondents listed spreading the word and making people aware of F/OSS as an activity within communities.

- **Lurking:** Several participants stated that they just read what others have written, either to learn or from general interest, but do not normally feel the need to participate in the discussions.

- **For fun:** A common theme in many of the responses was that members enjoyed being part of the F/OSS community. Many respondents wrote that they participated simply for fun.

While the first section of the survey identified the reasons that participants may have for participating in F/OSS communities and what they actually do, the second section was more concerned with how they are involved. The question put to them was how they make use of community forums. The suggested answers covered the issues of frequency of use, degree of involvement, and the effect of personal gain on involvement. This section was split into two distinct parts, the first exclusively looking at aspects of the community which are based around support provision and the second at the software development aspects. Only those participants who had experience of software development were asked to fill in the second part of this section.

Participants were presented with the following statements and asked to specify which of them best described their interaction with F/OSS community support forums:

- I usually do not use forums.
- I read what others have said but rarely participate myself.
- I sometimes participate but only when it's useful for me to do so.
- I often participate to help both myself and others.
- I often participate primarily to be social.

For this section, participants were asked to choose a single answer but were still given the opportunity to provide their own answer if none of the provided ones were appropriate. They were also invited to leave additional comments. Figure 2 shows the selections for this question. The most popular choice was "I often participate to help both myself and others," with 146 (33.7%) of respondents selecting it. Second most popular, with 127 (29.3%) selections, was "I read what others have said but rarely participate myself," followed closely by "I sometimes participate but only when it's useful for me to do so," with 109 (25.1%) participants selecting this.

There were a significant number of additional comments left for this question, the majority of which concerned the members' preference for e-mail based mailing lists over Web-based discussion boards. Several respondents stated that Web-based

Figure 2. Use of support forums

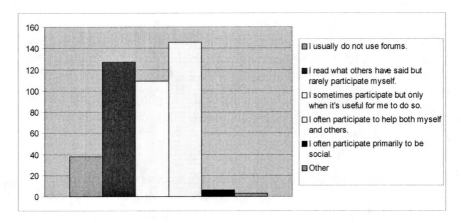

discussion boards were more suitable for beginners or "newbies." One of the main complaints of the discussion board was that the interface was awkward to use and not very efficient. Many participants also prefered the use of mailing lists because, unlike discussion boards, there is not the requirement to visit a Web site and log in using a pre-created account. The convenience of the mailing list is something that appears to be popular with many F/OSS community participants. Another viewpoint was that discussion boards were better for finding information on a specific topic, while mailing lists were better for asking questions and getting answers.

In addition to the mailing list comments, other comments were collected and are presented here in generalised categories:

- **Unwillingness to participate:** Many respondents felt uncomfortable or lacked the confidence to post on forums due to either lack of knowledge or an uncertaincy that what they are saying would be correct or useful.

- **Community building:** The importance of community building is something that many of the respondents clearly understood. Many wrote that they often participated to encourage others to participate and to build the community spirit by starting debates.

- **Giving back to the community:** Frequent remarks were written on the issue of giving back to the community. Some participants wrote that if one takes from F/OSS communities, then it is neccesary to give something back. Other comments were that F/OSS support communities are often the only way that a novice user or developer can give something back.

- **Passing on experience:** From the views of the respondents, experience certainly seems to count. Relating to the first point, many members wrote that they would only participate if they had experience with the issue under discussion. Several members also stated that as they received help from the community, their experience and knowledge grew, thus allowing them to provide more help to others.

- **Participation dependent on specific case:** Although the question attempted to obtain the general view of the use of community forums, several respondents stated that they could not generalise and that their experiences of F/OSS community varied greatly depending on the community.

The final section of the survey again addressed the issue of how members participate with communities but this time exclusively for the purpose of software development. Only those who had experience of F/OSS development were asked to complete this section of the survey. Although the survey effectively treated users and developers separately, it is acknowledged that developers are also users. This is not to say that all developers use the software that they have developed, although this may be the most common scenario (Gacek & Arief, 2004; Sowe et al., 2006). In an organisational context where developers are working on software projects, they may have no use for the software themselves. It is only those developers, working on software for their own needs that are likely to be users of their own software, that is, to "scratch a developer's personal itch" (Raymond, 2000). The real issue here is that both users and developers will make use of support forums but that only developers will make use of development forums. It is also accepted in some cases, particularly in the case of small-scale software projects, that support and development activities may take place in the same forums. Even so, it is still possible to differentiate between the two activities.

The following answers were provided and, as before, members could specify their own answers and leave additional comments:

- I mainly participate just to get help with my own development work.
- I participate both to receive help myself with my own work and to help others with theirs.
- I mainly participate to get involved in the development projects of others.
- I mainly participate to be sociable.

Out of the 434 respondents, analysis showed that 195 of them (45%) were involved with software development. Out of the 195 submissions, 123 (63.1%) had selected "I participate both to receive help myself with my own work and to help others with

Figure 3. Use of development forums

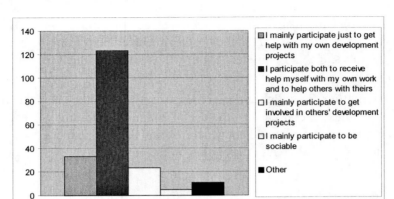

theirs" as their answer. This option had the highest number of selections by quite a large margin (see Figure 3). The second highest was "I mainly participate just to get help with my own development projects," with 33 selections (16.9%). This was slightly higher than the "I mainly participate to get involved in others' development projects" option which received 23 selections (11.8%).

Many additional comments were also left on the subject of the use of development forums. Again, only developers were asked to leave additional comments.

- **Contribution to documentation:** As well as contributing to code development, many developers also contributed to documentation on many projects. One developer wrote that by helping to document a piece of software they could learn more about the software itself.

- **Scratching the itch:** Many developers wrote that they really only contribute to either fix problems or add functionality to software that they needed. The problems often seem to be simple ones which, for instance, may prevent successful compilation of a piece of software. One developer wrote that people with similar itches would group together to work on a specific problem.

- **Professional development:** In some cases developers worked on and started F/OSS projects so they could learn about software development and pick up project management skills.

- **Paid development:** A relatively small number of developers wrote that their job involved F/OSS development. Clearly in these cases the motivating issues will be somewhat different than cases where the development is done by volunteers.

- **Project dependent:** It was the view of some that the developer's motivation to contribute and the type of contribution depend entirely on the individual project.

- **Derived work:** One respondent wrote that many F/OSS developers, especially those with limited experience, will produce software that it heavily based on other people's work. They also wrote that such software is often "disposable," i.e., that it would often have a very limited user base and be discarded after its use.

- **For fun:** As with all the survey questions, many developers responded by stating that they were involved in F/OSS development just for the enjoyment factor.

Research Analysis

F/OSS communities and the development that takes place within them are, by their very nature, open to anyone. If a person wishes to get involved in a project at almost any level and assume almost any role, they are free to do so. With F/OSS, anyone with the appropriate skills or the willingness to learn can be a software documentation writer, a developer, a tester, and so forth. This unrestrictive characteristic also applies to other aspects of F/OSS. Due to the expansion of the Internet into such a large number of countries across the world, F/OSS communities are not confined to any geographical constraints and exist on an international scale. These facts are justification for the use of the LUGs as the sample set for the research. Even though a LUG is usually based in a particular country and region, its members may be located anywhere in the world. Furthermore, each LUG member will more than likely be a member of many other F/OSS communities, depending on their software interests and needs. This is apparent from the comments left by many survey respondents who referred to other communities or software projects taking place outside their LUG. The empirical data collection conducted for this research is a very good example of the globally disparate and inter-networked existence of F/OSS communities. The request for survey participation was sent to five countries, that is, the United Kingdom, the United States of America, Canada, Germany, and Italy, as these were the countries with the highest number of LUGs. However, responses were received from Australia, China, France, India, Japan, Poland, and Spain. This made a total of twelve countries, only five of which had been targeted.

A potential limitation of the research which must be acknowledged lies in the differences between LUGs and other types of F/OSS communities. It is perhaps more likely that LUGs will be more orientated towards providing support than software development. This is because LUGs, unlike most other communities, are not usually based around a particular piece of software, and consequently may focus less on such development. Although most LUG members will be involved in other communities, there is no way of knowing the extent of external involvement from these results. Hence, it cannot be said for certain that the results of this research are entirely representative of F/OSS communities in general.

It is also acknowledged that through the use of the survey technique, the research has only reached those community members who are not averse to completing surveys. This is a general research problem that is very difficult to overcome without the use of multiple methods. However, as an individual's negative views towards the completion of surveys do not directly relate to the topics addressed by this research, this should not affect the results except by reducing the number of participants.

This research has looked at F/OSS communities using three perspectives, focusing on support, software development, and social interaction as separate activities. The results have come from different types of members who assume different roles within their communities. Nevertheless, significant trends become apparent upon analysis of the results. In general, community members perceive problem solving as their primary reason for participating with F/OSS communities. This result seems to concur with the literature, particularly Lakhani and von Hippel (2003), van Wendel de Joode, de Bruijn, and van Eeten (2003), and von Krogh, Haeflinger, and Spaeth (2003), and shows that people normally visit F/OSS communities when they have a problem and require help in fixing it. Interestingly, however, providing support to others was perceived as being almost as important. Providing support is also listed in much of the motivation literature, such as Lakhani and von Hippel (2003) and Sowe et al. (2006). The results suggest that community members perceive giving help and receiving it at almost the same level of importance and they are aware of the importance of sharing. Richard Stallman's ideology, emphasising the moral implications of sharing software, seems to be just as applicable now as it was during the early period after the creation of free software (Stallman, 1999). It is possible that members who receive more support than they give may be less inclined to fill in a survey. Nevertheless, the significant number of people who selected "providing support" as a reason demonstrates that the sharing ethic is very important amongst F/OSS community members.

Significantly fewer members had peer review of software and software development itself as reasons for participating with F/OSS communities. This was an expected result, as not all members of a community will be involved in software development. This is especially true of LUGs, which may not be involved in software development at all. However, the survey dealt with the respondents' general perceptions of F/OSS communities, not just the LUGs. Analysis of the responses showed that approximately 45% of respondents were involved with software development activities, including bug reporting, review, and so forth. Although it cannot be said for certain that this figure applies to F/OSS communities in general, it is interesting nonetheless. Peer review and software development received an almost equal number of selections for the first section of the survey. This would suggest that participants perceived them as being of almost equal importance. However, as the survey question suggested, peer review can be performed by members who have little or no knowledge of programming. They may instead simply test the software, suggest improvements or report bugs (Feller & Fitzgerald, 2002; Moody, 2001; Pavlicek, 2000; Raymond, 1999; Sowe et al., 2006; van Wendel de Joode et al., 2003). As not all respondents were developers, the pool of people involved with peer review may not be the same as those involved with software development and therefore the percentage of people who chose these options may not be comparable. However, upon further analysis, 127 respondents (29.3%) chose both peer review and software development, 65 respondents (15.0%) chose peer review and not software development and 51 (11.8%) chose software development and not peer review. Considering the selections both holistically and individually, it seems fair to conclude that peer review and software development are conducted by a similar number of participants and it could therefore be argued that they are perceived as being of similar importance. This also highlights the important relationship between user and developer (Feller & Fitzgerald, 2002; Scacchi et al., 2006). The user plays a much more important part in the F/OSS development process than in proprietary development, where their role is very limited.

One very interesting finding from the survey was the respondents' views on the social activities within F/OSS communities. When asked for their reasons for participating in F/OSS communities, 309 members (71.2%) selected "social exchange," defined as "To meet people or talk to people with similar interests." This number is slightly lower than the "problem solving" and "providing support" options which had 85.7% and 79.0% of selections respectively (see Figure 1). This clearly shows the apparent perceived importance of social exchange within these communities and would seem to fit the findings of Ghosh, Glott, Krieger, and Robles (2002), Hars and Ou (2001), Hertel et al. (2003), O'Mahony (2004), and Sagers (2004). However, when asked about their support-related activities, only 1.4% of members stated that they participated for social reasons. Likewise, only 1.2% wrote that they participated in software development activities for social reasons. This suggests that although social activities are very important and the community members perceive them as such, when it comes to software support and development the social aspects are seen as being of very little consequence. The work by Nonaka and Konno (1998) describes the concept of socialisation. This is the process of transferring tacit knowledge, which is intangible knowledge that cannot easily be recorded, between people through social interaction. Nonaka and Konno refer to socialisation as taking place in a face-to-face environment. However, if we apply the idea to F/OSS communities, it seems likely that the social interaction that takes place within support or development activities could facilitate the transfer and dissemination of tacit knowledge. Although the participants may not be aware of this, it would certainly be a very important process for the learning of a F/OSS community participant.

For example, consider how much knowledge can be encapsulated in a section of code. While the knowledge is not explicitly recorded there, the reader could learn a significant amount about the programming style and techniques used by the writer, and through discussion this tacit knowledge could be disseminated. It is therefore suggested that, although community participants may not perceive social activities as being important when it comes to software support and development, they are in fact using the social platform of community to transfer tacit knowledge in a manner similar to the socialisation process.

Phase two of the research investigated how members use communities for support purposes. This phase has produced some interesting results and suggests that people approach support activities within F/OSS communities differently. The results from the research's sample set revealed that the majority of members (33.7%) use the support aspects of communities to receive help and to help other people with their problems (see Figure 2). Although it may be possible that some members log in to a F/OSS community Web site with the sole intention of trying to help other users, it seems more likely that they will simply read the forums and should they come across a question they can answer, then they will do so. One comment left by a respondent stated that they often encounter requests for help, which they consequently reply to, while looking for help themselves. Although 33.7% chose this "receiving help and helping others option," 25.1% said they would only participate when it was useful for them to do so. It is probable that these members require some kind of incentive before they participate. It may be that they will only help solve problems that they themselves are encountering or that they feel they will only participate if they are sure they know the correct solution, so as to not damage their reputation. Issues identified in the literature of social reward, community identification and reputation, and reciprocity may also explain these findings (Hertel et al., 2003; Lakhani & von Hippel, 2003; O'Mahony & Ferraro, 2004; van Wendel de Joode et al., 2003). Whatever the case, some members seem to need more encouragement to participate than others.

Another significant finding from this phase related to the frequency of member participations, is that 29.3% of members in the sample, stated that they "read what others have said but rarely participate themselves." This is a large section of a community's members that will infrequently participate and will therefore have very little presence in the community. Zhang and Storck's (2001) research into "peripheral members" supports this finding. This may also be a matter of incentive or it may relate to the attitudes of the individual. It also seems probable that members who do not wish to participate in a F/OSS community forum will be unlikely to complete a survey. Consequently, there could be many more "lurkers" or members who rarely participate than the results of this survey have shown. From the results of this phase, a give-and-take attitude seems to be prevalent, concurring with the results of the first phase which showed that giving support and receiving support were perceived as almost equally important. It is also true, however, that F/OSS community members have different attitudes towards participation.

The third phase of the research focused on software development and provided a much more clear-cut set of results: 63.0% of developers who completed the survey stated that they participate with F/OSS community development activities "to get help with their own work, and to help others with theirs." There was a large gap between this answer and the second most popular, "I mainly participate just to get help with my own development projects," which was chosen by 16.9% of developers (see Figure 3). This convincingly demonstrates that most developers have a very collaborative attitude towards F/OSS communities. Reciprocity is a common theme in the literature, with studies showing that participants felt they were obliged to contribute because of feelings of loyalty, or because they wanted others to help in a similar way (Kollock, 1999; Lakhani & Wolf, 2003; Moon & Sproull, 2000; O'Mahony & Ferraro, 2004; von Krogh et al., 2003). These findings mirror the findings of the first two phases of the research and again show that the collaborative nature and sharing ethic is strong.

An issue which seems to have been deemed important by the participants of this survey is how F/OSS communities deal with the various ability levels of their members. Many comments were left concerning the differentiation between novices or "newbies," and experienced members. This divide seems to apply to both support and development aspects of F/OSS communities. While novices may sometimes be discouraged from participating due to a feeling of inadequacy, it is the cycle of learning and sharing knowledge that allows them to become experienced. Experience is what counts in these communities and as novices become experts, they can then help others to increase their knowledge and make the same transition.

The findings of this research are corroborated by the results of the FLOSS Final Report (Ghosh et al., 2002), which showed that learning and developing new skills, and sharing knowledge and skills, were the two most important reasons for a developer to join or stay in a F/OSS community. Sharing knowledge and skills was also shown to be the prime expectation of community members. The use of F/OSS communities as knowledge and skill sharing forums has been identified as an important reason for participation by many authors (Feller & Fitzgerald, 2002; Ghosh et al., 2002; Hars & Ou, 2001; Hertel et al., 2003; Lakhani & Wolf, 2003; Scacchi et al., 2006; von Krogh et al., 2003). Both research projects have clearly demonstrated that the giving and receiving ethos is at the heart of F/OSS community and that members are aware of this. It is therefore not true to say that F/OSS communities are chaotic and function with no systemic organisation.

Conclusion

The research presented here has collected data on the perceptions of F/OSS communities from the members that make them. This unique perspective has generated some fascinating findings about how community participants see themselves and their interaction and collaboration with the community. It has examined the ways that the members perceive the community that they belong to, and has shed more light on some of the motivational issues that are very specific to the individual member. The results have corroborated many of the findings from the literature on motivation and have also identified some issues that have not been properly addressed.

F/OSS communities are still seen by many outsiders as being chaotic, disorganised, and ad-hoc in nature. This research has shown that it is common interest, belief in ideals, and community relations that bind community members and communities together. The social activities are perceived as very important by members and help to keep a community together. The collaborative spirit applies to all areas of F/OSS, from support provision to software development. The qualitative data collected has also shown that F/OSS communities are not just confined to support and development, but are also host to many other activities, including the promotion of F/OSS, business and personal development, and educational activities. Furthermore, it is not just source code that is openly shared but also knowledge, skills, ideas, and expertise. It is the transference and dissemination of these, between a diverse set of participants, which allows F/OSS communities to function so effectively.

As F/OSS research is becoming more widespread, more avenues for investigation are opening up. There is much about F/OSS communities that is still unknown and more research is certainly needed. The details of knowledge and skill sharing within communities are still unclear as are the technicalities of the F/OSS development model. There is also much that can be learned from comparing other subject areas to F/OSS. For example, when one compares F/OSS with traditional software development and engineering, knowledge management, and work on the social dynamics of community, several parallels emerge that could contribute to our understanding of how F/OSS works. Likewise, an examination of similarities and differences between F/OSS and other types of online communities would also be beneficial. It is certain that, as F/OSS proliferates and becomes more and more a part of the information systems and computing world, so too will the research continue to develop and attempt to explain the F/OSS phenomenon.

References

Bartle, R. (2003). *Designing virtual worlds.* IN: New Riders Publishing.

Bezroukov, N. (1999). *Open source software development as a special type of academic research (critique of Vulgar Raymondism). First Monday.* Retrieved July 2006, from http://www.firstmonday.dk/issues/issue4_10/bezroukov/index.html

Dibona, C., Ockham, S., & Stone, M. (Eds.). (1999). *Open sources: Voices from the open source revolution.* CA: O'Reilly & Associates, Inc.

Ducheneaut, N., Yee, N., Nickell, E., & Moore, R. (2006). "Alone together?" Exploring the social dynamics of massively multiplayer online games. *ACM Conference on Computer Human Interaction (CHI 2006)*, Montréal, Canada.

Edwards, K. (2001). *Epistemic communities, situated learning and open source software development.* Retrieved May 2006, from http://opensource.mit.edu/papers/kasperedwards-ec.pdf

Elliott, M. (2005). *The virtual organisation culture of a free software development community.* Retrieved March 2005, from http://opensource.mit.edu

Feller, J., & Fitzgerald, B. (2000). A framework analysis of the open source software development paradigm, *International Conference of Information Systems*, Brisbane, Australia.

Feller, J., & Fitzgerald, B. (2002). *Understanding open source software development.* Addison Wesley.

Gacek, C., & Arief, B. (2004). The many meanings of open source. *IEEE Software*.

Gattiker, U. E. (2001). *The Internet as a diverse community: Cultural, organizational, and political issues.* NJ: Lawrence Erlbaum Associates, Inc.

Ghosh, R. A., Glott, R., Krieger, B., & Robles, G. (2002). *Survey of developers, free/libre and open source software: Study and survey*, International Institute of Infonomics, University of Maastricht, The Netherlands, Retrieved May 2006, from http://www.infonomics.nl/FLOSS/report/

Hann, I. H., Roberts, J., & Slaughter, S. A. (2004). Why developers participate in open source software projects: An empirical study. *Twenty-Fifth International Conference on Information Systems*.

Hars, A., & Ou, S. (2001). Working for free? Motivations of participating in open source projects. *Thirty-fourth Hawaii International Conference on System Sciences*, Hawaii.

Hertel, G., Niedner, S., & Herrmann, S. (2003). Motivation of software developers in open source projects: An Internet-based survey of contributors to the Linux kernel. *Research Policy (Special Issue on Open Source Software Development)*. Retrieved February 2004, from http://opensource.mit.edu/papers/hertel.pdf

Hildreth, P., Kimble, C., & Wright, P. (1998). Computer mediated communications and international communities of practice. In *Proceedings of Ethicomp'98*, The Netherlands.

Kollock, P. (1999). The economies of online cooperation: Gifts and public goods in cyberspace. In M. Smith & P. Kollock (Eds.), *Communities in cyberspace*. London: Routledge.

Krishnamurthy, S. (2002). *Cave or community? An empirical examination of 100 mature open source projects*. Retrieved February 2005, from http://opensource. mit.edu

Lakhani, K. R., & von Hippel, E. (2003). How open source software works: 'Free' user-to-user assistance. *Research Policy No. 32*. Elsevier Science.

Lakhani, K. R., & Wolf, R. G. (2003). *Why hackers do what they do: Understanding motivation effort in free open source software projects* (Working paper). MIT Sloan School of Management. Retrieved February 2004, from http://freesoft-ware.mit.edu/papers/lakhaniwolf.pdf

Lanzara, G. F., & Morner, M. (2003, October 2). The knowledge ecology of open-source software projects. Paper presented at seminar on *ICTs in the Contemporary World*, LSE Department of Information Systems.

Lerner, J., & Tirole, J. (2002). Some simple economics of open source. *Journal of Industrial Economics, 50*, 197-234.

McKenna, K., & Bargh, J. (2000). Plan 9 from cyberspace: The implications of the Internet for personality and social psychology. *Personality and Social Psychology Review, 4*, 57-75.

Moody, G. (2001). *Rebel code: How Linus Torvalds, Linux and the open source movement are outmastering Microsoft*. UK: The Penguin Press.

Moon, J. Y, & Sproull, L. (2000). Essence of distributed work: The case of the Linux kernel. *First Monday*. Retrieved July 2006, from http://firstmonday. org/issues/issue5_11/moon/index.html

Nonaka, I., & Konno, N. (1998). The concept of 'Ba': Building a foundation for knowledge creation. *California Management Review, 40*.

O'Mahony, S., & Ferraro, F. (2004). *Hacking alome? The effects of online and of-fline participation on open source community leadership*. Retrieved December 2004, from http://opensource.mit.edu

Oh, W., & Jeon, S. (2004). Membership dynamics and network stability in the open-source community: The ising perspective. *Twenty-Fifth International Conference on Information Systems*.

Pavlicek, R. C. (2000). *Embracing insanity: Open source software development*. Sams Publishing.

Raymond, E. S. (1999). *The magic cauldron*. Retrieved December 2003, from http://www.catb.org/~esr/writings/cathedral-bazaar/magic-cauldron/

Raymond, E. S. (2000). *The cathedral and the bazaar*. Retrieved November 2003, from http://www.catb.org/~esr/writings/cathedral-bazaar/cathedral-bazaar

Rheingold, H. (2000). *The virtual community*. MIT Press.

Ribeiro, G. L. (1997). Transnational virtual community? Exploring implications for culture, power and language. *Organization Science, 4*(4).

Sagers, G. W. (2004). The influence of network governance factors on success in open source software development projects. *Twenty-Fifth International Conference on Information Systems*.

Scacchi, W., Feller, J., Fitzgerald, B., Hissam, S., & Lakhani, K. (2006). *Understanding free/open source software development processes*. Retrieved December 2005, from http://www.ics.uci.edu/~wscacchi/Papers/New/SPIP-FOSS-Intro-Dec2005.pdf

Schoberth, T., Preece, J., & Heinzl, A. (2003). Online communities: A longitudinal analysis of communication activities. *Hawai'i International Conference on System Sciences*, Hawaii.

Schofield, A., & Mitra, A. (2003). Free and open source software communities as a support mechanism. *UK Academy of Information Systems conference 2005*, Newcastle, UK.

Smith, M. A., & Kollock, P. (1994). *Managing the virtual commons: Cooperation and conflict in computer communities*. American Sociological Association.

Smith, M. A., & Kollock, P. (1999). *Communities in cyberspace*. London: Routledge

Sowe, S., & Stamelos, I. (2005). Identification of knowledge brokers in OSS projects mailing lists through social and collaborative networks. *Tenth Panhellenic Conference on Informatics*, Volos, Greece.

Sowe, S., Stamelos, I., & Angelis, L. (2006). Identifying knowledge brokers that yield software engineering knowledge in OSS projects. *Information and Software Technology, 48*, 1025-1033.

Stallman, R. (2000) The GNU operating system and the free software movement. *In* C. DiBona, S. Ockman, & M. Stone (Eds), *Open sources: Voices from the open source revolution* Sebastopol, CA: O'Reilly & Associates.

Torvalds, L., & Diamond, D. (2001). *Just for fun: The story of an accidental revolutionary*. New York: Harper Business.

van Wendel de Joode, R., de Bruijn, J., & van Eeten, M. (2003). Protecting the virtual commons. In T. M. C. Asser (Ed.), *Information technology & law series*, 44-50.

von Krogh, G., Haeflinger, S., & Spaeth, S. (2003). *Collective action and communal resources in open source software development: The case of freenet.* Retrieved January 2005, from http://opensource.mit.edu

Walther, J. B. (1996). Computer-mediated communication: Impersonal, interpersonal and hyperpersonal interaction. *Communication Research, 23*, 3-43.

Wellman, B., & Gulia, M. (1995). *The reality of virtual communities.* American Sociological Association.

Yee, N. (2006). The demographics, motivations and derived experiences of users of massively-multiuser online graphical environments. *PRESENCE: Teleoperators and Virtual Environments, 15*, 309-329

Zhang, W., & Storck, J. (2001). Peripheral members in online communities. *Americas Conference on Information Systems*, Boston.

Section III

Tools for Qualitative F/OSS Development

Chapter V

Software Development and Coordination Tools in Open Source Communities

Ruben van Wendel de Joode, Twynstra Gudde Management Consultants, The Netherlands

Hans de Bruijn, Delft University of Technology, The Netherlands

Michel van Eeten, Delft University of Technology, The Netherlands

Abstract

Open source communities bring together a dispersed collection of people, sometimes a large number of them, around the development of open source software. In the absence of enforceable formal structures, like those found in corporate settings, how are the activities of all these participants coordinated? From the outside look-ing in, it may seem that chaos and disorder rule. It is true that most decisions are made on an individual basis by the participants themselves. Nevertheless, work is coordinated successfully. We first discuss mechanisms that reduce the need for co-ordination—most notably by striving for modularity and elegance. We then turn to a number of mechanisms that provide coordination even under the difficult conditions that are present in open source communities. We conclude by outlining a number of potential steps for future research.

Introduction

Open source software (OSS) communities are truly intriguing and fascinating phenomena. In OSS communities, the source code, which is the human-readable part of software, is treated as something that is open and that should be downloadable and modifiable to anyone who wishes to do so. The availability of the source code has enabled a practice of highly decentralized software development in which large numbers of people contribute time and effort. These large numbers of people are not confined to certain geographical places; on the contrary, they come from literally all continents. And yet despite the highly decentralized and geographically-dispersed development process, the software that is developed in some of the communities is of a high quality. The Apache and Linux software, for instance, are said to be of a high quality.[1] These are just some of the characteristics that explain why so many people find OSS communities absolutely fascinating.

Unsurprisingly, OSS communities are heavily researched. Until now, much of the research efforts have focused on an analysis of the individual contributors in OSS communities. Typically, the focus is on the questions: Who are the contributors in OSS communities? and Why are people motivated to contribute their time and effort in OSS communities? Currently, we are beginning to find answers to both questions. An extensive survey performed by the University of Maastricht, for instance, has provided some answers to the first question. It showed that, although the development process is global, most participants are concentrated in the United States of America and various European countries like France, Germany, Italy, and the United Kingdom. People were found to have differing backgrounds, motives, and skills. At the same time, most people in the survey reported to spend less than 10 hours a week in OSS communities (Ghosh & Glott, 2002). Also, answers have been found on the second question. Lakhani and von Hippel (2003), for instance, have argued that the costs of participation in the communities are relatively low. The finding of low costs implies that a low level of benefits may already provide sufficient motive to contribute. Also, research has shown that people are motivated through a wide range of differing benefits. Some of these benefits are: learning and improving one's programming skills (e.g., Hertel, Niedner, & Herrmann, 2003; von Hippel & von Krogh, 2003; Lakhani & Wolf, 2003); meeting a personal need with a software program that has a certain functionality (e.g., Edwards, 2001; Hars & Ou, 2002); and having fun (Lakhani & von Hippel, 2003; Torvalds & Diamond, 2001).

Although research on individual contributors is important, it is just one part of an explanation of what OSS communities are and how they "operate." A topic that is just as essential, or maybe even moreso, is *coordination*. To coordinate means more than just to getting individuals motivated; it also means to arrange and order individuals' efforts; it refers to labor division and to task specialization. In short, coordination refers to the core challenge of organizations (March & Simon, 1993).

This aspect of OSS communities will be the focus of this chapter. The related question that dominated the research leading up to the chapter is: How are the activities of individuals in open source communities coordinated? Some research in this respect has been performed (e.g., Garzarelli, 2003; von Hippel & von Krogh, 2003; Kogut & Metiu, 2001; Markus, Manville, & Agres, 2000; Narduzzo & Rossi, 2003), however, as of now we only have some partial answers.

There are a number of reasons why this chapter is relevant for practitioners and policymakers alike. The most important reason is that more and more organizations are switching or intending to switch to OSS. Also, in many countries policymakers are actively promoting the adoption of open source software. Many reasons are used to justify this choice, for instance, it is claimed that the costs of OSS are low, open source software stimulates the local ICT market, and OSS provides more freedom to users. It is, however, unclear what the implications are for organizations when they adopt OSS. How, for instance, should they interact with the developers in a community? To answer this and related questions we first need to have a better understanding of coordination in OSS communities.

Background: The State of the Art

A growing body of literature focuses on specific parts of OSS communities. Topics are, for instance, the role of users in processes of innovation in the communities (von Hippel, 2001; Shah, 2003), the role of modularity in both the community structure and the structure of the software (Narduzzo & Rossi, 2003), and how conflicts arise and are resolved (Elliot & Scacchi, 2002). Some of the mechanisms that have come out of previous research are listed below. The goal is not to be exhaustive, but rather to give an idea of the variety of mechanisms that have been identified. Furthermore, many of the previously identified mechanisms that are not included in the list will be discussed in the remainder of the chapter:

- **Automated mailing lists:** One of the basic tools to support coordination in open source communities is mailing lists. They are used to communicate all sorts of information to the participants in the communities (Bauer & Pizka, 2003; Edwards, 2001; Kogut & Metiu, 2001).

- **Software modularity:** Many researchers call attention to the fact that open source software is modular, which basically means that big and complex software programs are divided into smaller parts. These smaller parts are relatively easy to understand and ensure that programmers remain relatively less dependent on each other (Benkler, 2002; Bonaccorsi & Rossi, 2003; Garzarelli, 2003; Kogut & Metiu, 2001; Langlois, 2002; Lerner & Tirole, 2002,

p. 28; McKelvey, 2001b; Moon & Sproull, 2000; Narduzzo & Rossi, 2003; Tuomi, 2001).

- **Open source licenses:** Many communities have licensed their software with a specific type of software license. Examples of open source licenses are the General Public License (GPL) and the Berkeley Software Distribution (BSD) license. These licenses allow others to adopt and make amendments to the software and are said to be necessary to ensure collaboration in the communities (Boyle, 2003; Dalle & Jullien, 2003; Lerner & Tirole, 2002; McGowan, 2001; O'Mahony, 2003).

- **Versioning systems:** Many researchers have pointed to the relative importance of versioning systems and in particular the concurrent versions system (CVS) in achieving coordination in open source communities (Bauer & Pizka, 2003; German, 2002; Hemetsberger & Reinhardt, 2004; von Krogh, Haefliger, & Spaeth, 2003; Scacchi, 2004; Shaikh & Cornford, 2003). Versioning systems are systems that support the development activities of individuals and allow multiple developers to simultaneously improve a certain piece of source code. Adopting a versioning system thus reduces the need for coordination among the participants in a community.

Previous research has contributed a more detailed understanding of OSS communities and resulted in a number of mechanisms that are said to promote coordination. Yet, by and large the mechanisms are identified and discussed in isolation. In this chapter the mechanisms are jointly presented and new mechanisms are identified. Special attention will be given to the ways in which each mechanism contributes to coordination. Also, an overall picture of coordination will be sketched.

Methodology

The data presented in this chapter are based on an exploratory analysis of documents, Web sites and mailing lists, a study of secondary literature, and, in particular, face-to-face semi-structured interviews with experts who are actively involved in open source software development. Sixty people from the Netherlands, the US, and Germany were interviewed. All interviews were fully transcribed. A coding scheme is used to refer to the transcriptions of the interviews. In the chapter, references to the interviews have the following structure: [10XX]. Most interviewees were members of one of the following four OSS communities: Linux, Apache, PostgreSQL, and Python. The communities were selected based on their size, as measured by the number of contributors and users. This is important, since it means that the chapter is limited in its focus to communities that have attracted large numbers of people who are motivated to contribute. For this reason, some of the mechanisms and tools presented in this chapter might not be relevant to and not be present in the smaller OSS communities. The mechanisms and tools should not be taken to explain how and why communities attract new contributors and users.

Individualism Dominates the Creation of Open Source Software

We know that developers are mutually dependent and must collaborate to develop a complex product like software (Hans de Bruijn, 2002; Hans de Bruijn & ten Heuvelhof, 2000). However, many interviewees claim to develop and maintain the software in a very individualistic way, based on their individual choices and preferences:

Well, you have a pool of people who do what they want to do. Nobody gives me an assignment. Instead, I think, 'hey how weird, this process is very slow.' Well, then I myself will start to work to solve it ... This is of course a very selfish approach, but I think that most people work like that. They work on what they run into. (R1032)

Developers decide for themselves what they find interesting and what they would like to work on. McCormick (2003) writes that the general principle in OSS communities is that "decisions are made 'by those who are willing to do the work' and that in the long run, developers tend to support the best technical solution for a problem" (p. 31).

The fact that open source developers behave individualistically has three consequences. First, participants in the communities have no way of ensuring that other participants will perform the activities they are asked to perform. Participants work independently and no one can force anyone to do anything. A respondent claimed that few free software developers work together. "A company like Microsoft hires developers, puts them in a room and makes them work together. We don't because we do not have walls ... We cannot make them do anything, we can only suggest" (R1008).

The second consequence is that things only get done when someone wants to do them. "That is actually what the entire open source philosophy is about. Things only get done if at least one person feels that they are important. That person makes sure that it works" (R1008). In other words, an activity will not be performed unless someone in the community considers it important.

The third consequence of individualism is that participants in the community keep limited oversight of what others are doing. "We try and keep things simple. We like to spend more time getting stuff done instead of tracking everything" (R1019). The developers in the communities can be compared to ants in "a big ant colony. You don't know what someone else is doing" (R1059).

Somehow, individuals are able to create software in a highly individualistic manner that nevertheless works remarkably well. This notion adds quite some complexity to the question that is central in this chapter, namely: How are the activities of individuals in open source communities coordinated?

The remainder of the chapter will introduce and discuss a great number of mechanisms that are argued to relieve the need for formal planning and centralized coordination, i.e., allow a very decentralized and individualistic development process.

Mechanisms to Relieve Some of the Need for Collaboration and Coordination

If software becomes complex and the interdependencies increase, the need for collaboration and coordination increases. In such a situation, changes to one part of a program would likely have the effect that "something else does not work anymore" (R1049) and therefore the software-development process "is no longer controllable" (R1049). To prevent these problems and to relieve some of the need for rules and structuring activities, developers in OSS communities have two guiding principles. These allow the number of lines of source code to grow without proportionally increasing the overhead needed to coordinate, as the mechanisms reduce the burden of coordination. The mechanisms are *elegance* and *modularity*.

Elegance of Software: Making the Software Easier to Understand

Donald Knuth is often seen as the founder of the concept of elegant code (Wallace, 1999). Elegance is basically a technical characteristic of software:

Combining simplicity, power and a certain ineffable grace of design ... The French aviator, adventurer, and author Antoine de Saint-Exupéry, probably best known for his classic children's book The Little Prince, was also an aircraft designer. He gave us perhaps the best definition of engineering elegance when he said "A designer knows he has achieved perfection not when there is nothing left to add, but when there is nothing left to take away."[2]

According to Ellen Ullman, elegance is structured and reductive. It is a term used to indicate that software works and, at the same time, a notion of beauty. "So from the standpoint of a small group of engineers, you are striving for something that is structured and lovely in its structuredness" (Rosenberg, 1997).

Elegance is sometimes claimed to be an indisputable characteristic of source code. Source code is either elegant or it is not. The more experienced and skilled software programmers are claimed to be best judges of whether source code is elegant. "If you narrow your circle to a small group of good developers then it is easy to decide what software is elegant and what is not" (R1032).

From an organizational perspective, elegance performs a role in enabling coordination and collaboration in social groups. Compared to inelegant source code, elegant code is effectively easier to understand and likely to better express "what it is doing while you are reading it" (R1018). It implies that the software will not be doing something "in a non-intuitive, ineffective way" (R1018). The fact that elegant source code is easier to understand has a number of advantages. First, it is easier to make changes to it. "You can only work when the software is beautiful and elegant, to be able to implement changes easily" (R1032). Thus, although the number of lines of source code is bound to increase when the functionality of software is enriched, an elegantly written piece of software provides some counterforce to complexity and to a certain degree ensures that the code remains relatively easy to understand and to change.

Another advantage of elegance is that programmers have to invest less time and effort to understand what the source code is trying to accomplish. From reading elegant source code, programmers claim to be able to understand what the source code aims to achieve and to judge whether the source code indeed accomplishes the task. The relative ease with which developers can understand a certain piece of code effectively lowers the time needed to decide whether a certain piece of source code is good. This enables them to make decisions without paying more attention to reading and understanding source code than is strictly needed. Elegance, for instance, enables maintainers or others in the community to decide whether a patch should be included. It also allows people who were not previously involved in the community to improve code without spending much time and effort deciphering the source code.

Elegance also relieves the need for coordination and collaboration. Because the code is elegant, it is easy to understand what the effects of a change in one part of the software will be for other parts. Elegance allows developers to either adjust other parts of the software or to ask others to take a look at it. If the source code were inelegant, the chance is much higher that a minor change in one aspect would have dramatic effects for others. In that case, developers must collaborate, simply because they cannot oversee the consequences of their changes. They need the help of others to discover and solve the mutual dependencies in the software.

Modularity: Untangling the Complexity of Software Programs

The second mechanism, modularity, is based on the idea of "divide and conquer" (Dafermos, 2001). Modular software is divided into smaller pieces, building blocks, which together form a software program. Consider the following statement from a Linux developer: "Of course, Linux software is very complex ... The answer is to divide and conquer. When you have a complex piece of software, you cut it into ten pieces and if you manage to provide them with good interfaces then you only need to understand the separate pieces." (R1032)

Thus, modularity depends on two aspects, namely, the modules and the interfaces that connect the modules. Essentially, the modules are the different parts of a software program that perform separate tasks and which can act independently of one another. Software that is divided into modules has the big advantage that each module performs a limited set of activities or tasks, which "individuals can tackle independently from other tasks" (Lerner & Tirole, 2002, p. 28). An ASF board member explained how the Apache software came to be an entirely modular software program and described the advantages of this modularity:

In 1995 Robert Tow rewrote the entire NCSA server to make it entirely modular. The server became much easier to maintain. You could work on part of the server, without having to worry that you would damage the rest ... [Developers] could work in parallel without stepping on each other's toes. (R1008)

Thus, modularity is said to reduce the costs of coordination (Benkler, 2002; Bonaccorsi & Rossi, 2003; Dafermos, 2001; Garzarelli, 2003; Kogut & Metiu, 2001; Lerner & Tirole, 2002, p. 28; Narduzzo & Rossi, 2003).

For modularity to work, however, the interfaces between modules must be well defined and changes to the interfaces must be kept to a minimum. Modularity allows a developer to work independently "as long as she gets the communication interfaces right" (Weber, 2004, p. 173). An interviewee explained, "Free software maintains complexity with such a loose structure because the interfaces are well defined" (R1018). Hence, clear interfaces are important to ensure that modules can successfully be integrated into one software program. The need for clearly-defined interfaces that remain relatively unchanged leads to at least one question: How are open source developers who base their decisions on their individual preferences capable of keeping the interfaces of software modules constant? Isn't it likely that a developer will want to make changes to a module that requires a change in the interface?

This is not the case according to one of the interviewees. He argues that OSS developers tend to be "almost perverse" (R1018) in their strive to implement development perfectly along the standard interfaces. Does this imply that developers build difficult and unnatural solutions to a problem? If so, do developers value modularity above elegance? According to another interviewee this is also untrue. He said that modules can always be changed in ways that have no effect on either the elegance of the code in the module or the interface that connects the module to other modules. "If you have a glass of beer and drink out the rim, then I know how to model my mouth and lips. If the content under it changes but the rim stays the same I would still know how to hold my mouth. So the code or technique doesn't matter, as long as it fits in with the interface" (R1009).

One can, however, seriously doubt the obviousness that is suggested in the quote. Innovative solutions can demand changes in existing interfaces. In that case, developers must somehow coordinate their changes to ensure that the modules remain compatible. Thus, modularity does relieve some of the need for coordination, but it also results in trade-offs with, for instance, innovation.

Mechanisms to Coordinate Massive Amount of Individual Efforts

Participants in open source communities spend little time coordinating their activities. For instance, they do not send e-mails explaining to other developers that they are currently developing software module X or translating software program Y. Generally, developers do, however, use a large and rather sophisticated infrastructure that supports their activities and, in the process, coordinates their efforts with those being invested by hundreds if not thousands of other developers. As communities start to grow and attract more and more participants, they tend to support their activities with an increasingly sophisticated technical infrastructure. "These drastic changes were possible, mainly because of the technical conditions that have improved" (Bauer & Pizka, 2003, p. 172). This infrastructure consists of mechanisms that computerize coordination. Next to the technical infrastructure, participants have adopted a number of devices that are related to a specific way of working. Many communities have further adopted methodologies and standards to coordinate their individual efforts.

Table 1. Coordination and collaboration tools

Coordination Tools
Versioning systems
Project leadership
Mailing lists
Bug-tracking system
Coding style guides
To-do lists
Added text
Names attached to improvements
Small and incremental patches

The next pages describe the mechanisms and explain for each mechanism (1) what it is, (2) what it does, (3) how it leads to the coordination of individual efforts, and (4) why it allows participants to spend as much of their time as possible on the actual development and maintenance of source code.

Versioning System

Most open source communities support their development and maintenance activities with a software versioning system. Well-known examples of such systems are the concurrent versioning system (CVS), subversion (SVN), and bitkeeper. These systems are automated systems that allow remote access to the source code and they enable multiple developers to work on the same version of the source code simultaneously. According to a number of interviewees, one of the advantages of using a versioning system is that it becomes unimportant to know whether participating developers are able to write high-quality source code, since they cannot cause major damage to the source code. Older versions of the source code are automatically stored in the system and can be used as a backup. An ASF director explained, "It actually doesn't matter whether the committers are any good. In a CVS you cannot cause a lot of damage to the software. First, the older versions of the software are saved and you can return to these versions" (R1008).

Basically anybody can download source code from the versioning systems. Some communities, however, have restricted access to their system. In these restricted communities, only participants with *committer status* can upload source code. Developers without committer status must send their patch to someone in the community who has that status. The way in which people gain committer status depends on the community. In most communities if you write and submit a few good patches of source code you will soon receive an invitation to become a committer: "To become a committer is really easy. You can become a committer with only one e-mail." [R1008] The same interviewee explained how their versioning system supports the software-development process in the Apache community:

In a CVS you start with a new version. Then something gets added and a new version is created. Then again something new is added. This way the line grows longer and longer. This line is called the "head": the most recent version. When someone disagrees with something, he can take an earlier version from the line and implement his changes to that version ... For every new version you have to explain what you changed, why and in which way. You write this information in a log, which enables other developers to understand what you have done. (R1008)

Much literature has addressed the importance of versioning systems to support software development and maintenance (German, 2002; Hemetsberger & Reinhardt, 2004; von Krogh et al., 2003; Scacchi, 2004; Shaikh & Cornford, 2003). Basically, versioning systems support the decentralized development process (Himanen, 2001) in a number of ways. First, participants can access the versioning system simultaneously. They do not have to wait until another developer has finished working on the source code. Second, the presence of logs is important. The log files provide participants with an explanation of how the source code works and what it intends to accomplish. Third, the versioning systems allow participants to move back in the development line and take an older version of the source code. This enables them to take out a commit that at a later stage of development proves to be bad code. This last option effectively reduces the need for participants in the community to monitor and analyze the value of every new commit. It could also be an explanation as to why most communities make it fairly easy to become a committer.

Project Leadership

Many communities have one or more clearly identifiable project leaders (Bonaccorsi & Rossi, 2003; Dafermos, 2001; Fielding, 1999; Hann, Roberts, Slaughter, & Fielding, 2002; Lerner & Tirole, 2002; Markus et al., 2000; McCormick, 2003; Moon & Sproull, 2000). Generally, leadership is "given" to the person who makes the first lines of source code publicly available. Other communities, like Apache, have leadership vested in a board of directors. The community members periodically elect a new board of directors. The actual activities performed by project leaders and boards of directors are different for each community. For instance, the project leader might maintain the version that is commonly accepted as the official version of the software, as well as promoting the software and informing organizations considering adopting the software about how to deal with the larger community of developers. Project leaders and directors thus perform an important role in coordinating the efforts of the participants (Egyedi & van Wendel de Joode, 2004).

Probably the best-known project leader is Linus Torvalds in the Linux community. Much research has acknowledged the leadership role of Torvalds in the Linux community. It claims that the development process is centralized and that Torvalds is the one with decision authority (Kogut & Metiu, 2001; McGowan, 2001; McKelvey, 2001a; Moon & Sproull, 2000). Torvalds, however, does not manage the entire community alone. So-called "lieutenants" have authority over subsystems of the Linux kernel. Torvalds does have a number of sources of power. To start with, he is the trademark owner of the name Linux. He also owns the collective copyright of the Linux kernel. Finally, he maintains the most popular and most used version of the Linux kernel, referred to as the "standard version."

Although the importance of project leaders is identified by many, there is no universally accepted image among researchers of their degree of influence on the individual contributors. von Hippel and von Krogh (2003), for instance, argue that project leaders are different from most managers in "traditional" companies, because they cannot enforce (p. 218). They argue that further research on the role of project leaders is highly relevant, and propose that project leaders should perhaps be compared to coaches in a sports team.

One of the better analyses of leadership is that done by McCormick (2003). He conducted many interviews with community members to understand the role of project leadership and ascertain whether they have more influence than other individuals in the community. The conclusion is that they do have more influence. The question is, "What can they do with this control?" To which the answer is, "Nothing much." "Unless you are paying someone, you can't do much of anything to officially delegate tasks. All you can do is cheerleading, or simply debating some ideas until the other person gets excited about it" (McCormick, 2003, p. 11). If the project leaders try to impose too much control and exercise power, this "would be resisted by the group" (McCormick, 2003, p. 27).

To summarize, project leadership is an important part in ensuring coordination among the contributors, as they perform many activities that benefit the joint development of the software. However, their actual influence over individual contributors is rather limited.

Mailing Lists

The presence and use of mailing lists is another part of the infrastructure that supports collaboration (Bauer & Pizka, 2003; Edwards, 2001; Kogut & Metiu,

2001). Every community has a number of mailing lists on which different issues are discussed. These lists serve different purposes and target different audiences. Certain lists focus on users, providing them with a forum to ask questions and receive answers. One respondent we spoke to considered these mailing lists to be one of the most important tools for supporting the development of software in OSS communities, because "people on these lists dare to say things and really want to hold their ground" (R1032). In short, the mailing lists provide the developers with a forum to exchange and discuss their ideas, and they also give users a forum to ask questions and receive answers.

Bug-Tracking Systems

Bugs are basically mistakes or flaws in a software program. Many software bugs are discovered while the software is actually in use. Users of open source software often write a "bug report" when they come across a mistake or when they find that something does not work (e.g., von Hippel, 2001). The challenge of a bug report is to precisely describe the problem the user encountered. Such a report involves a description of, for instance, the kind of hardware being used, the other software installed on the computer, the particular configuration of the software, and the error message encountered.

To manage the processes of writing a bug report, solving a bug, and communicating the solution to fix the bug, communities have created and adopted different systems. "Sometimes bug reports are handled through mailing lists. Other projects use bug-tracking systems" (R1009). More advanced bug-tracking systems, for instance, have a format in which a bug report should be written. Such a format includes a number of questions that need to be answered. The report of the bug is then stored in the system, where it awaits someone to fix the bug.

Effectively, the more advanced systems eliminate the need for people in the communities, first, to contact others and explain about the bug they found and, second, to convince another developer to solve the bug. Instead, the bug-tracking system allows people anywhere in the world to report and describe a bug whenever they like. Also, anyone can access the repository of bug descriptions and can decide to fix a bug that has not yet been solved.

Coding Style Guides

Communities like Apache and Linux have *coding style guides* (Egyedi & van Wendel de Joode, 2004; Kroah-Hartman, 2002). The coding style guides prescribe how a piece of source code should be styled. Each community makes its own choices about the style and layout of the source code.

Using different styles in one software program makes the program difficult to understand. One interviewee told how another programmer "changed the entire source code to the GNU style. Although, this is just a minor change, it does make it almost impossible to compare the programs, because the indentation is completely different" (R1032).

Thus, the coding styles aim to achieve unified definitions and a single style of writing software among all developers in a community. This uniformity or standardization reduces the time needed to understand source code written by other participants and diminishes the need to communicate about a piece of software. It thus increases independency in the communities and allows for a more decentralized development effort.

To-Do Lists

Participants typically have many ideas about how a software program should work or what new features should be added to a program. The only way these ideas are transformed into actual lines of source code is by someone writing the source code. The problem is that not everyone is able to do so, for instance, because they lack the skills and time or because they are simply not motivated to do the job. Yet the ideas might nonetheless be valuable and very much needed by a large part of the community.

To ensure that new ideas are actually transformed into source code, participants have created to-do lists. As the name implies, a to-do list is an inventory of things that at least one person considers important or wants to have. One community that has adopted a to-do list is the PostgreSQL community. Typically, an idea is transferred to this list when someone sends "an e-mail with a suggestion" (R1016).

The to-do list is a coordinative mechanism because it signals developers as to what others in the community find important. Participants do not have to discuss and explain why they find the ideas important. Instead, they just put the item on the to-do list. Others can take a look at the list and judge for themselves what they find interesting and what they would like to work on. As such, the to-do list serves as a marketplace in which demand, for example, for a certain feature, meets supply, namely participants who have the knowledge, time, and motivation to develop the feature. To-do lists are another example of a mechanism created and adopted with the goal of structuring the efforts of individuals in the communities.

Added Text: An Important Signaling Function

Many participants in open source communities try to write source code that is elegant. Sometimes, however, they cannot. Take for instance software in which a security hole is discovered. In such a situation, the first and foremost goal is to fix the hole to ensure that the software is secure again. Whether the code of the solution is elegant is less important (R1011). Another example in which elegance is less important is where developers write software that is compatible with proprietary software. Typically, the participants have no access to the actual source code of that software and therefore they have to guess how the source code looks based on the way the software functions. To write software that is compatible with proprietary software is very difficult and frequently results in inelegant code. "Sometimes code has to look inelegant to work around something. Most of the time we see curse words accompanying these postings, explaining why it is not elegant. It is inelegant because it has to deal with other [proprietary] software"(R1018).

Thus, in some situations it is almost impossible for developers to write elegant code. Yet when other developers read the inelegant code, they have a hard time understanding how the software works and why that particular piece of source code is needed. They may even be tempted to remove the code or think that the author is a poor programmer. Why else would someone have written such inelegant code?

Interviewees indicated two general ways for an author to communicate the presence of inelegant code and the fact that circumstances forced them to write the inelegant code. One way is the use of curse words written in the sidelines of the source code. These curse words signal three things to other participants who read the source code. First, the curse words communicate that the author of the code knows that the code is inelegant as written, but that no other way has as yet been found to work around a certain problem. Second, the curse words explain what the source code accomplishes. Hence, the curse words try to reduce the time that a participant needs to understand the inelegant source code and they are a way to make it as easy as possible for others to understand and possibly modify the code. Third, the number of curse words serves as an indication of the elegance of the code and thus provides a warning to others that the source code is considered to be inelegant and thus difficult to understand. One interviewee remarked, "In some architectures you see more curse words than in others. If you take a piece of code for the IBM mainframe, there you would see no curse words. In code for the Spark on the other hand you see a lot of curse words ... because the IBM mainframe is elegant" (R1018).

A second way to indicate the presence of inelegant code is by placing the inelegant code between brackets and surrounding the code with #ifdef. "This makes it easier to delete and replace the code with an elegant solution in the future" (R1054). Adding this text to the code performs a similar function to the curse words added in the sidelines of source code.

Adding text, like curse words or #ifdef, not only communicates the inability of the developer to write an elegant solution for a certain problem. It also has a clear function in neutralizing some of the negative effects of inelegant code, namely helping other developers to understand the code and modify it.

Names Attached to Improvements

In every open source community, one or more mechanisms are adopted to relate participants to their contributions. One such mechanism is the credits list, which contains the names of developers who have contributed to the development of software. The credits list also specifies what they have contributed. "My name is attached to every KDE program" (R1059) In most programs the name of the person that fixed a part of the software is put "next to the resolved item" (R1016). The Apache community "makes a point of recognizing all contributors on its Web site, even those who simply identify a problem without proposing a solution" (Lerner & Tirole, 2002, p. 27). For many participants, having their names attached to the contributions is an important acknowledgement of their work and part of the reason why they contribute their time and effort. "I am actually quite proud to see my name on a piece of code" (R1055).

Another way in which a contributor is related to a certain piece of source code is the voting system that is adopted in the Apache community. For every new commit[3] a vote is held. "Don't expect too much from [the vote], if no one is against the patch then the patch will remain in. If you explicitly voted that you wanted the patch to remain then you commit yourself to help clean up the software if the patch turns out to be less good" (R1008)

Connecting participants to their contributions also fulfills a coordinative function. The contributor of the source code is known and thus feels responsible. "If something turns out to be wrong it is my mistake. Usually the one who made the last change is the most appropriate person" (R1056) to fix the problem. If a user of the source code discovers that it does not work, for whatever reason, then the user knows whom to e-mail. Chances are, the contributor will feel responsible, if only because the contributor's name is attached to the source code that does not work. Typically, the contributor fixes the problem. "Iterations work like this; someone will comment on a patch and the author will make the changes and re-submit the patch. It rarely happens that someone ignores the comments. If this happens and it is valuable to me I would go in and do it myself ..." (R1019).

Small and Incremental Patches

Another coordinative mechanism is the norm that developers should only contribute source code in small and incremental patches. Keeping the patches small is important, as it makes it easier for others to understand what the source code aims to achieve and how it intends to do so. "This system has too many improvements. I have to be able to understand it. Please, only send small patches" (R1056). An example is an incident in the Apache community in which the developers rejected a patch from a company because "it was simply too big. The developers [in the Apache community] could not unwrap it to understand how it worked" (R1015).

Thus, small and incremental patches make it easier for others to understand what the source code aims to achieve. This is because (1) they lower the amount of time participants need to invest to understand what the source code intends to do; (2) they make it easier for other developers to make modifications to the patch; and (3) they lower the value of resources that are destroyed when a new addition is not accepted or is removed. This last point can be explained as follows. A large patch of source code can be viewed as a collection of small patches. Consider a participant who combines the small patches into one large patch and adds this source code into the code base. One small mistake in the collection of patches could compromise the integrity of the entire program. In this case, participants might have to remove the entire collection of patches. If the patches had been contributed in small pieces, the participants could remove just the small patch of source code that created the problem. In that case the destruction of value, that is, the time and effort spent to create the rejected patch, would have been relatively small, as it involved just one change. In general, the bigger the patch, the more time and effort is involved in developing and maintaining the software. For these reasons, Linus Torvalds is unlikely to add large patches. "I will repeat my rule: I do not apply large patches with many separate changes. I apply patches that do one thing."[4]

To Summarize: What is Behind All This?

The long list of mechanisms forces us to think about the question: What can we learn from all this? What can we learn from the mechanisms about the way in which coordination is achieved in the communities? One thing that is very clear is that the mechanisms are relatively straightforward and easy to understand. The role of mechanisms like the bug-report system and the versioning system are clear and understandable. The mechanisms are easy to copy and, as a matter of fact, similar mechanisms can be found in other disciplines and in other settings.

Although the mechanisms are understandable, the mechanisms do indicate the presence of an organizational structure that is more encompassing; a structure that is surprisingly intelligent. A small detour: Survey research shows that many individuals participate in open source communities to achieve their own, individual goals. For instance, they want to create software that solves their problems ("to scratch an itch"), they want to achieve status and reputation, and they want to learn. Furthermore, a growing number of companies participate and use open source software to earn a profit. Yet, although the actions of many individuals and companies are based on individualistic goals and interests, their actions do benefit the collective. For instance, a developer who solves his own individual problem is likely to help others, as there is a very real chance that others will face similar problems. Thus, solving the problem of oneself will frequently help others in the collective. It is here, on the interface between the collective and the individual level, where we see the important role of the mechanisms. The mechanisms allow individuals to make their own choices and perform the activities they want to do, but at the same time the mechanisms ensure that almost all of these activities contribute to the development and maintenance of software. In other words, the mechanisms transform self-interested actions into actions that benefit the collective.

This characteristic—the fact that self-interested and individualistic behavior aggregates into collective processes that benefit all—is possibly the most fascinating conceptual characteristic of open source communities. However, it also has practical implications, as it implies that the communities that were analyzed in this research and the software-development processes in them can hardly be directed and/or controlled. One could argue that in some communities a project leader or board of directors has power to control the software development process. Yet, we have also seen that their actual influence over individual activities is rather limited. Hence, the conclusion that the development of software in the communities is ad-hoc; it is a patchwork of contributions received from many different individuals. It also implies that it is very difficult to actually design open source communities. The mechanisms presented in this chapter can be copied to other settings and they can provide valuable lessons to other organizations. However, it does not mean that creating an infrastructure based on the mechanisms is sufficient to create a flourishing community in which new and innovative software is developed.

Future Trends

This chapter lists the findings from an explorative study on OSS communities focusing on the ways in which developers and users coordinate their activities to create software. The main result of the chapter is a list of mechanisms and tools that support the processes of collaboration. The mechanisms that are listed in the chapter were found in at least one of the four communities that were analyzed in light of our research. The identified mechanisms are highly valuable and important to understand OSS communities. Ideally, however, it is just a first step in our search to understand the actual workings of the communities. There are five next steps that could follow this research and would hugely contribute to the state-of-the-art knowledge on OSS communities. They are, in no particular order, summarized below.

First, some of the mechanisms, like modularity and elegance, create trade-offs. To write software that is extremely modular means that people can work in a decentralized fashion and relatively autonomously. To write modular software, however, will also create trade-offs. Modularity, and especially keeping the interfaces intact, might, for instance, conflict with goals like speed of processes, memory management, or security of the software. The goal of achieving extremely modular software might even be in conflict with the wish to write elegant software. How do developers in the communities make such trade-offs? Is modularity always the most important goal to strive for? What are the reasons developers choose other goals and values above modularity or elegance? Future research might shed light on these trade-offs, which could add crucial insight in the inner workings of OSS communities.

Second, the identified mechanisms are not necessarily a guarantee for the success of a community. The mechanisms were found in at least one of the communities that were analyzed in this research. The criteria for selecting the communities were, among other criteria, based on the number of contributions and the lines of source code. The communities that were selected in this research needed to have a high score on both these measures. After having identified the mechanisms, a logical next question would be to understand the relationship between each of the mechanisms and the success of a community. Is it true that successful communities have a minimal number of coordination mechanisms in place? What are these mechanisms? And if successful communities have adopted similar mechanisms, are these mechanisms adopted because of the success of a community, or is the success of a community the result of adopting a number of crucial mechanisms? A better understanding of the relationship between community success and coordination mechanisms could contribute to the search of the do's and don'ts in the design of OSS communities.

Third, what is the relative importance of each of the mechanisms? The chance that some mechanisms are more important than others to achieve coordination is quite high. Quantitative research could shed a light on the relative importance of each of the mechanisms, which, again, would lead to crucial insight in the do's and don'ts in designing OSS communities.

Fourth, a highly relevant line of research would focus on efficiency. The focus in this chapter is on the mechanisms as such. It does not include an analysis about the efficiency of a mechanism or an analysis about the extent to which a certain mechanism contributes to the overall efficiency of the software-development process. Do the mechanisms increase the overall efficiency? What norms are in use for a given mechanism and how do they affect the efficiency of the mechanism? This line of research would be highly relevant for developers in open source communities and for companies that are thinking about adopting certain mechanisms within their corporate software-development processes.

Finally, more and more companies are becoming involved in OSS communities. As a result, a growing number of developers are being paid by companies to participate in the development processes. Relatively new communities, like Eclipse, have been deliberately created and managed by companies. In Eclipse, almost all developers are paid by a limited number of companies, and the developers are instructed to develop the software in certain directions. These communities are different from the communities that are analyzed in this research. The coordination processes in a community like Eclipse are likely to differ from the coordination processes in communities like Linux and Python. A promising line of research would be to understand these differences: What are the differences and what are the consequences of these differences on for instance the quality of the software and the level of efficiency in the communities?

Conclusion

The goal in this chapter is to understand how activities in open source communities are coordinated. It was argued that individualism dominates most of the activities. People spend little time to formally inform each other every time they start a new activity or continue on someone else's work. Generally speaking, they ask no permission to start a project or consult others to determine whether they believe a certain project is valuable. Instead, they simply *do*. They decide, for themselves and based on their own reasons, to create or rewrite software, for instance, to update software documentation, or to remove a bug. As a result, a general overview of who does what and when is lacking.

From the outside looking in on the communities, such a lack might be mistaken for chaos and disorder. Or, in the terms of one of our interviewees who explained coordination in the Linux community: "Linux is anarchy! Everybody does what he feels like doing." This research, however, shows that the apparent disorder and chaos is misleading and besides the truth. In reality, the active participants in open source communities use a great number of mechanisms and tools that structure their individual activities and, at the same time, allow for a highly decentralized development process. Mechanisms like bug-report systems, versioning systems, and to-do lists cost little time and effort for people to use, but they result in the coordination of their activities. They ensure that reported bugs are visible and can easily be found by those who want to fix bugs. There is no direct need for a bug reporter to communicate with the person fixing the bug; they operate independently of each other. Thus, mechanisms like a bug-report system effectively bring structure and coordination to a community of people that to outsiders might give the impression of being highly disorganized.

The collection of mechanisms and the development process provide an important lesson to organizations: To increase the chance that their contributions are accepted and their wishes listened to, organizations should use the mechanisms that are available and that are being used by others in a given OSS community. Furthermore, the mechanisms and the individual approach to software development demonstrate a general inability for anyone to actually enforce preferences upon the community as a whole. Developers and organizations must develop software they have a need for. They cannot enforce their preferences on other developers.

Acknowledgment

The authors would like to thank Tineke M. Egyedi (Delft University of Technology), Alessandro Rossi (University of Trento), and the two anonymous reviewers for their valuable comments and suggestions on previous versions of this chapter. Also, we would like to extend our gratitude to all the open source and free software developers who contributed to this research. The standard disclaimer applies.

References

Bauer, A., & Pizka, M. (2003). The contribution of free software to software evolution. Paper presented at the *Sixth International Workshop on Principles of Software Evolution (IWPSE'03)*, Helsinki, Finland.

Benkler, Y. (2002). Coase's penguin, or, Linux and the nature of the firm. *Yale Law Journal, 112*(3), 369-446.

Bonaccorsi, A., & Rossi, C. (2003). Why open source software can succeed. *Research Policy, 32*(7), 1243-1258.

Boyle, J. (2003). The second enclosure movement and the construction of the public domain. *Law and Contemporary Problems, 66*(1/2), 33-74.

de Bruijn, H. (2002). *Managing performance in the public sector*. London; New York: Routledge.

de Bruijn, H., & ten Heuvelhof, E. (2000). Networks and decision making (1st ed.). Utrecht: LEMMA Publishers.

Dafermos, G. N. (2001). Management and virtual decentralised networks: The Linux project. *First Monday*. Peer reviewed journal on the Internet, *6*(11). Retrieved December 2004, fromhttp://www.firstmonday.dk/issues/issue6_11/dafermos/

Dalle, J.-M., & Jullien, N. (2003). "Libre" software: Turning fads into institutions? *Research policy, 32*(1), 1-11.

Edwards, K. (2001). Epistemic communities, situated learning and open source software development. Paper presented at the *Epistemic Cultures and the Practice of Interdisciplinarity Workshop* at NTNU, Trondheim.

Egyedi, T. M., & van Wendel de Joode, R. (2004). Standardization and other co-ordination mechanisms in open source software. *Journal of IT Standards & Standardization Research, 2*(2), 1-17.

Elliot, M. S., & Scacchi, W. (2002). *Communicating and mitigating conflict in open source software development projects.* Unpublished research paper, Institute for Software Research, UC Irvine. Retrieved December 2004, from www.ics. uci.edu/~melliott/commossd.pdf

Fielding, R. T. (1999). Shared leadership in the Apache project. *Communications of the Association for Computing, 42*(4), 42/43.

Garzarelli, G. (2003). Open source software and the economics of organization. In J. Birner & P. Garrouste (Eds.), *Austrian perspectives on the new economy.* London: Routledge.

German, D. M. (2002). The evolution of the GNOME project. Paper presented at the *Meeting Challenges and Surviving Success: 2nd Workshop on Open Source Software Engineering, 24th International Conference on Software Engineering*, Orlando, USA.

Ghosh, R. A., & Glott, R. (2002). Free/libre and open source software: Survey and study. Maastricht: MERIT, University of Maastricht.

Hann, I.-H., Roberts, J., Slaughter, S., & Fielding, R. T. (2002). Why do developers contribute to open source projects? First evidence of ecomic incentives. Paper presented at the *Meeting Challenges and Surviving Success: 2nd Workshop on Open Source Software Engineering, 24th International Conference on Software Engineering*, Orlando, USA.

Hars, A., & Ou, S. (2002). Working for free? Motivations for participating in open-source projects. *International Journal of Electronic Commerce, 6*(3), 25-39.

Hemetsberger, A., & Reinhardt, C. (2004). Sharing and creating knowledge in open-source communities: The case of KDE. Paper presented at the *Fifth European Conference on Organizational Knowledge, Learning, and Capabilities*, Innsbruck, Austria.

Hertel, G., Niedner, S., & Herrmann, S. (2003). Motivation of software developers in open source projects: An Internet-based survey of contributors to the Linux kernel. *Research Policy, 32*(7), 1159-1177.

Himanen, P. (2001). The hacker ethic and the spirit of the information age. New York: Random House.

von Hippel, E. (2001). Innovation by user communities: Learning from open-source software. *Sloan Management Review, 42*(4), 82-86.

von Hippel, E., & von Krogh, G. (2003). Open source software and "private-collective" innovation model: Issues for organization science. *Organization Science, 14*(2), 209-223.

Kogut, B., & Metiu, A. (2001). Open-source software development and distributed innovation. *Oxford Review of Economic Policy, 17*(2), 248-264.

Kroah-Hartman, G. (2002). Proper Linux kernel coding style. *Linux Journal,* July 1.

von Krogh, G., Haefliger, S., & Spaeth, S. (2003). *Collective action and communal resources in open source software development: The case of freenet.* Unpublished manuscript, research paper, University of St. Gallen, Switzerland. Retrieved November 2004, fromopensource.mit.edu/papers/vonkroghhaefligerspaeth.pdf

Lakhani, K., & von Hippel, E. (2003). How open source software works: "Free" user-to-user assistance. *Research Policy, 32*(7), 922-943.

Lakhani, K., & Wolf, R. G. (2003). Why hackers do what they do: Understanding motivation and effort in free/open source software projects (Working Paper no.4425-03). Boston: MIT Sloan. Retrieved November, 2004 fromfreesoftware.mit.edu/papers/lakhaniwolf.pdf

Langlois, R. N. (2002). Modularity in technology and organization. *Journal of Economic Behavior and Organization, 49*(2), 19-37.

Lerner, J., & Tirole, J. (2002). Some simple economics of open source. *Journal of Industrial Economics, 50*(2), 197-234.

Markus, M. L., Manville, B., & Agres, C. E. (2000). What makes a virtual organization work? *Sloan Management Review, 42*(1), 13-26.

McCormick, C. (2003). *The big project that never ends: Role and task negotiation within an emerging occupational community.* Unpublished recent summary of findings, as part of PhD research, University of Albany. Retrieved December 2002, from opensource.mit.edu

McGowan, D. (2001). Legal implications of open-source software. *University of Illinois Review, 241*(1), 241-304.

McKelvey, M. (2001a). The economic dynamics of software: Three competing business models exemplified through Microsoft, Netscape and Linux. *Economics of Innovation and New Technology, 10,* 199-236.

McKelvey, M. (2001b). *Internet entrepreneurship: Linux and the dynamics of open source software.* Unpublished working paper, Centre for Research on Innovation and Competition, The University of Manchester, Manchester. Retrieved December 2004,from http://les1.man.ac.uk/cric/Pdfs/DP44.pdf

Moon, J. Y., & Sproull, L. (2000). Essence of distributed work: The case of the Linux kernel. *First Monday, 5*(11).

Narduzzo, A., & Rossi, A. (2003). *Modularity in action: GNU/Linux and free/open source software development model unleashed.* Unpublished working paper,

Quaderno DISA n. 78. Retrieved December 2004, from opensource.mit.edu/papers/narduzzorossi.pdf

O'Mahony, S. C. (2003). Guarding the commons: How community managed software projects protect their work. *Research Policy, 32*(7), 1179-1198.

Rosenberg, S. (1997). *Elegance & entropy: Ellen Ulman talks about what makes programmers tick.* Retrieved October 9, 1997 from Salon.com

Scacchi, W. (2004). Free/open source software development practices in the computer game community. *IEEE Software, 21*(1), 56-66.

Shah, S. (2003). Com*munity-based innovation & product development: Findings from open source software and consumer sporting goods.* Unpublished dissertation, MA Institute of Technology, Cambridge, MA.

Shaikh, M., & Cornford, T. (2003). *Version control software for knowledge sharing, innovation and learning in OS.* Paper presented at the Open Source Software Movements and Communities workshop, ICCT, Amsterdam.

Torvalds, L., & Diamond, D. (2001). Gewoon voor de fun (Just for Fun) (C. Jongeneel, Trans.). Uithoorn: Karakter Uitgevers.

Tuomi, I. (2001). Internet, innovation, and open source: Actors in the network. *First Monday, 6*(1).

Wallace, M. (1999). *The art of Don E. Knuth.* Retrieved September 16, 1999, from salon.com

Weber, S. (2004). *The success of open source.* Cambridge: Harvard University Press.

Endnotes

[1] Based on http://www.infoworld.com/article/03/07/01/HNreasoning_1.html (March, 2004) and http://www.reasoning.com/newsevents/pr/02_11_03.html (August, 2004).

[2] Cited from the jargon file http://www.retrologic.com/jargon/E/elegant.html (October, 2005).

[3] A commit in the Apache community is the act of adding source code to the versioning system.

[4] Cited from the Internet http://kt.zork.net/kernel-traffic/kt19991101_41.html (September, 2003). The role of Linus Torvalds in the community and his ability to "enforce" a rule on other contributors in the Linux community has been addressed in more detail in a previous section of this chapter.

Chapter VI

Evidence-Based Assurance to Support Process Quality in the F/OSS Community

Anas Tawileh, Cardiff University, UK

Omer Rana, Cardiff University, UK

Wendy Ivins, Cardiff University, UK

Stephen McIntosh, Cardiff University, UK

Abstract

This chapter investigates the quality issues of the free and open source software (F/OSS) development processes. It argues that software developed within the F/OSS paradigm has witnessed substantial growth rates within the software developers' community. However, end users from outside the community are still sceptical about adopting F/OSS because of the perceived lack of quality assurance mechanisms within the F/OSS development process. The authors aim to promote higher adoption of F/OSS artefacts outside the developers' community by exploring possibilities to provide appropriate evidence based assurances that F/OSS artefacts will meet the quality levels expected by users.

Introduction

The F/OSS paradigm introduces methodologies and development models different from those usually utilised within the proprietary software industry, based on cooperation and collaboration among developers from all over the world. It spans geographical and cultural boundaries more than any other human community or business enterprise. A particular project may therefore involve contributions from a large community of developers. However, the potential benefits associated with this diversity come at a price, as the donated code can be of varying quality (Stamelos et al., 2002). Whilst the F/OSS community has developed novel mechanisms to effectively tackle some of the difficulties encountered during its endeavours (particularly those associated with code interoperability), quality concerns remain a largely unexplored area (Glass, 2001).

End users usually argue that the F/OSS community is very technically oriented (Lerner & Tirole, 2002). Although it is claimed to be widely open for participation, it has a distinct vocabulary and a high level of requisite knowledge for access, which renders the low entry boundaries effectively useless. Because of these perceptions, users from outside the community are highly sceptical about adopting and considering F/OSS as a viable alternative to proprietary software (Feller & Fitzgerald, 2000). User scepticism about the community has resulted in adoption levels much lower than those that could be expected. However, this could be rectified if there were clear comparison criteria based on the merits of the benefits realised by the user in adopting the F/OSS option compared to the use of proprietary software.

Concerns about quality, and subsequent trust in the software developed through the F/OSS process remain a significant limitation to adoption. This could be attributed to the relative failure of the F/OSS community in promoting the quality of its artefacts. Although members of the community may argue that the quality levels attainable within the F/OSS community are much higher than those of proprietary software, these claims will not have any significant value unless supported by evidence-based assurances and communicated effectively to users from outside the community (Michlmayr, Hunt, & Probert, 2005). The community could overcome this issue by developing appropriate mechanisms to signal quality to end users in simple, easy to understand terms. These need to assure users that the produced artefacts will satisfy their needs and requirements in order to encourage users to embrace a more open attitude towards the adoption and reliance on F/OSS artefacts. Therefore, we argue that it is necessary to provide metrics for validating the quality of the F/OSS artefacts, which themselves must follow the open source concept, and be verifiable using tests that are accessible (and potentially repeatable) by end users.

This chapter aims to investigate the possibility of solving this problem by developing appropriate mechanisms to assure end users that artefacts developed within the F/OSS community will fulfil their needs and satisfy their expectations in order to encourage wider adoption of F/OSS outside the developers' community. The chapter considers modelling of the overall F/OSS development process using soft systems methodology (Checkland, 1999) and UML, and relates to particular metrics that can be associated with critical points along this development process. We also describe how such metrics could be used to improve the development process, and demonstrate how automated extraction of such metrics could be employed alongside F/OSS repositories such as Sourceforge.net. An illustrative example is used to demonstrate the practical applicability of the proposed approach, with a description of an early prototype of the supporting information system.

Background

During the past few years, adoption of F/OSS witnessed a striking worldwide increase (Wheeler, 2006). This growth was attributed to different factors that encouraged users to favor F/OSS over proprietary software, such as cost savings, performance, and security. However, levels of adoption differ between different organisations. While organisations and individuals who were involved in the development of F/OSS software are observed to have high levels of reliance on F/OSS, others who had less experience with the development process are still relatively sceptical about embracing this software (Feller & Fitzgerald, 2000).

The development models that evolved within the F/OSS community proved to be highly effective in managing complex, highly distributed projects and facilitating communication and collaboration among developers in a very diverse geographical and cultural environment (O'Reilly, 1999). F/OSS development models were based on the ideas of intensive communication between developers, large dependency on peer reviews, and frequent release of source code. Developers believed that reliance on these ideas would accelerate bug discovery and fix release time, and substantially enhance the quality of the produced software (Raymond, 2000).

These beliefs are perfectly sufficient for developers to convince themselves of the levels of quality and reliability they can expect from software developed within the F/OSS community. However, this view of quality is based on the perception that all users of software are developers. This assumption is very far from being true. Although availability of source code could facilitate the discovery of software bugs and enable checking the claimed quality of the application, non-developers should not be expected to have the requisite time and knowledge to pursue such activities (Fuggetta, 2003). It is also often hard to fully understand the impact of version changes of particular library routines on a particular application. Non-developers usually focus on evidence-based quality and measurable performance metrics that could indicate the level of trust to be placed on the software, without going through the usually exhaustive task of performing code-level inspections. It is therefore necessary to extend the availability of source code with specific evidence-based metrics that could be used to evaluate a given F/OSS artefact.

Availability of source code is not enough to guarantee quality of software to users other than developers. Bernard Golden (2004) argues that "Source code access is not usually enough for pragmatic IT organizations," he claims that even when the software source code is available, there remains a question about who created that code. Golden suggests that F/OSS products contain code that was written by many different contributors, and that users have questions about the trustworthiness of the source code that results from these contributions.

Another factor that magnifies the suspicious attitudes of users (particularly within commercial organisations) against F/OSS is the absence of a clear mechanism for user involvement, particularly in the requirements gathering phase. Ideally, software is created to address the needs of a particular user or group of users (Raymond, 2000). The goal of maximising financial profits and ensuring the long-term survivability of the company motivates commercial software development firms to interact with users, understand their needs and provide them with application software that fulfils those needs. In the case of F/OSS, users are developers themselves, and they usually initiate software development projects according to their own requirements. There is no formal user participation in the decision-making process and no formal requirements elicitation phase (Bartkowiak, 2004). Consequently, the vast majority of software developed within the F/OSS community in its early days was applications targeted towards a highly experienced user base, and paid less attention to user friendliness (Nichols & Twidale, 2003). Most of these were server side and network applications, with a clear shortage of desktop and end user software.

The F/OSS community noticed this issue, and attempted to bridge the apparent gap between technically aware and lay end users. These efforts were primarily stimulated by the increasing commercial interest in F/OSS and the proliferation of businesses operating on the basis of F/OSS software and services. The most obvious example to illustrate this development is the evolution of the GNU/Linux operating system. In its early days, the installation and user interface of the operating system were mainly command-line based and relatively difficult to use, especially for a non-technical user. However, the community realised that in order to gain a reasonable share of the desktop operating systems market, the development process must understand the users' needs and respond by providing them with the features they desired. Coupled with the support of commercial F/OSS development companies, the GNU/Linux operating system evolved through rapid development cycles until it became much easier to use and provided features that the lay end user could utilise simply and effectively. Currently, desktop versions of the GNU/Linux operating system compete with the major proprietary counterparts on aspects of features and user friendliness.

In his book, *Crossing the Chasm*, Geoffrey Moore (1999) suggested a classification of technology users into three types: early adopters, pragmatists, and late adopters. So far, the majority of F/OSS software is still the area of early adopters, those who are comfortable with unfinished products, and have sufficient technical knowledge to solve problems as they occur (Ousterhout, 1999). However, most business organisations are usually less tolerant of failures and risks. Therefore, because software is an essential element in the operation of any commercial entity, they are not usually willing to undertake risk by adopting technology early, and they are mostly considered to be pragmatists. End users with less technical knowledge fall within the same category, as their main interest is to achieve their goals with minimum effort, and they always try to avoid the frustration they usually feel when a technical problem arises.

In order to overcome the barriers inhibiting the adoption of F/OSS by organisations and individuals with less technical knowledge than those within its community (the pragmatists), some mechanism should be developed to signal quality and reliability of artefacts developed within the community to outsiders. These assurances would permit users to evaluate the level of confidence they can place on the desired artefact without going through the exhausting process of analysing source code or delving into information contained within the artefacts' communication/documentation repositories. Bernard Golden (2004) describes this requirement as: "Technology providers must cross the chasm to sell pragmatists ... Open Source must cross the chasm as well."

Current Efforts

Some studies have been conducted to investigate the quality assurance processes and activities within the F/OSS community. Michlymar et al. (2005) examined the current status of quality in F/OSS projects, and identified different characteristics of the F/OSS development process that affect quality. Others (Zhao & Elbaum, 2002; Ioannis et al., 2002) surveyed and examined many F/OSS projects to evaluate the different claims made about the quality of F/OSS artefacts. Our work takes an additional step forward to suggest enhancements to the current F/OSS development process in order to improve the quality of its resulting artefacts.

Several initiatives were also proposed to assist the evaluation of F/OSS quality. Each approached the situation in a slightly different manner. Most of these initiatives suggest a formal methodology to assess the suitability and maturity of F/OSS artefacts for a particular use within an organisation. These methodologies are based on standardised analytical frameworks to compare multiple potential F/OSS artefacts in order to choose the most suitable option.

The open source maturity model (OSMM) was developed by Navica (2003) to aid organisations in determining the maturity of an F/OSS artefact (referred to as a "product"). It proposes the assessment of a product's maturity based on different product elements, such as product software, support, documentation, training, product integration, and professional services. The methodology suggests a three-phase process where the organisation conducts assessments for each of these product elements, followed by defining a weighting for each element based on the organisation's requirements, and finally calculating the product's overall maturity score.

CapGemini open source maturity model (Duijnhouwer & Widdows, 2003) is another initiative developed by CapGemini, an international technology consulting firm, to enable organisations to determine whether an open source artefact is suitable for their needs. The model suggests 27 indicators grouped within four groups: product, integration, use, and acceptance, upon which an overall score could be calculated to determine the maturity level of any open source artefact. The model also mandates a feedback phase where the effectiveness of the utilised indicators can be evaluated. It proposes that the continuous evaluation of such indicators would allow evaluation within a changing environment, which is particularly important for the development approach adopted in the F/OSS community.

Another recent initiative by Intel Corporation, the Centre for Open Source Investigation at Carnegie Mellon University and SpikeSource, proposes an open standard to facilitate assessment and adoption of F/OSS artefacts: the Business Readiness Rating (BRR) (2005). This initiative intends to give companies a trusted, unbiased source for determining whether the open source software they are considering is mature enough to adopt and help adopters assess which F/OSS software would be most suitable for their needs. The rating weighs the important factors for successful implementation of F/OSS software, and includes: functionality, quality, performance, support, community size, security, and others. Initial analysis suggests that this effort provides more detailed evaluation data and scoring than the previously mentioned models.

All the previous initiatives aim to provide a systematic methodology to assist adopters of F/OSS artefacts in assessing and evaluating the suitability of the different available artefacts according to their own requirements. Although they use slightly different methods and techniques, they all rely on the same underlying concept. All these methodologies attempt to evaluate the quality or maturity of F/OSS artefacts using information that is already available. They rely on factors such as number of releases, number of security vulnerabilities, reference deployments, market penetration, and others. Even though these factors provide very good indicators about the project's characteristics, maturity, and quality, and could provide valuable assistance in the evaluation process, these approaches have two major drawbacks:

1. Assessment of these factors is highly subjective. While some of the factors could be easily measured and assessed based on objective evidence, most cannot be measured or judged objectively. This leads to a wide margin of objectivity of assessment results which may be purposefully exploited to favour one artefact over another. The proposed models attempt to overcome this weakness using different approaches. BRR, for instance, emphasises the openness of their proposal as a guarantee for correct use of the model. CapGemini's Open Source Maturity Model proposes that various measures be determined by a panel of experts who have demonstrated knowledge of the field.

2. The proposed models attempt to evaluate F/OSS artefacts (as is), by gathering and analysing information that is available about the artefact. They do not offer any suggestions or measures to improve quality or maturity of the artefacts developed within the F/OSS community. Proponents of these models argue that the mere existence of a standard, such as an open evaluation methodology for F/OSS artefacts, is sufficient to stimulate competition within the community projects, which in turn will increase quality and maturity levels of the developed artefacts. However, this quality enhancement would be difficult to measure, as the underlying determining factors are highly subjective.

We attempt to address these weaknesses by investigating the relevant processes involved within the F/OSS community, to explore possible solutions that could enable the community to provide users with sufficient assurances that the artefacts it produces would function as intended. The aim of this investigation is to present end users with tools that could facilitate the selection of appropriate artefacts produced by the F/OSS community according to their specific needs and requirements. This could increase the willingness to adopt F/OSS artefacts outside the developers' community, and eventually enhance the quality of these artefacts. The analysis presented here takes a considerably different approach to the existing methodologies, in that it focuses on delivering evidence-based solutions that could be verified and validated by users themselves or any individual or organisation who might be interested in the F/OSS artefact in some objective manner.

Requirements

Defining requirements in an open context such as the F/OSS community is not an easy undertaking. The many interacting (and sometimes contradicting) technical, social, and ideological aspects magnify the difficulty. Also, the field of F/OSS is relatively new, and although the amount of research performed is increasing at a substantial rate, the number of available empirical studies is comparatively low.

To lay the foundation for the study, expectations of both F/OSS developers and end users should be clearly identified. The most frequent user arguments made when asked about their reluctance to adopt F/OSS can be summarised as follows:

1. There are a number of F/OSS products with varying levels of quality. How can we choose the appropriate product that would satisfy our needs among all the available different options (Golden, 2004).

2. Artefacts developed within the F/OSS community are built by volunteer contributors. How can we guarantee that this development model will not result in a chaos that would impair the quality of the developed artefacts (Jones, 2004).

3. Although F/OSS proponents claim that the availability of source code and the transparency of the development process should provide sufficient guarantee of the quality levels of the produced artefacts, the language used within the community is highly technical. Therefore, making sense of the communication and collaboration repositories available in each project requires specialised knowledge and expertise, needing time and resources to obtain that knowledge (Fuggetta, 2003; Golden, 2004).

When these questions are presented to the F/OSS community members, they usually argue that the natural evolution of the community resulted in a widely open, transparent development process with very low entry barriers. They claim that these characteristics provide sufficient assurances about the quality and reliability of artefacts developed. The large number of F/OSS projects is a healthy sign of competition, and the process has adequate built-in features that would minimise (or virtually eliminate) malicious contributions (Angelius, 2004).

The last point is particularly relevant, as developers insist that the best way to guarantee the quality of artefacts is to keep their source code largely open (Raymond, 2000). Anyone who may be interested can eyeball the code to ensure its quality and security and whether it would be suitable for their requirements. It cannot be denied that such examination would require substantial requisite knowledge that should be demonstrated by anyone willing to undertake such activity. The argument by the community suggests that the benefits obtained by using F/OSS artefacts should come at a price, which users must pay if they want to increase their confidence in the artefacts they want to adopt. Otherwise, they should (trust) the community and tolerate any associated risks. This reasoning ignores two important points: first, most end users, particularly businesses, fall under the "pragmatists" category (Moore, 1999). When faced with a critical decision of relying on software, these users usually tend to pick the least risky option. Unfortunately, trusting the F/OSS community (or maybe any community) is still perceived as a risky undertaking. Second, the increasingly larger sizes of F/OSS projects render the option of eyeballing every single line of code practically useless (Bartkowiak, 2004). This size and complexity is further magnified when users have to evaluate different projects to select the most suitable option.

The previous discussion clearly suggests the need for possible solutions to communicate the quality of artefacts developed within the F/OSS community to outsiders. Any proposed solution should take into consideration the applicability and acceptability by the community. Additionally, such solutions could encourage end users and businesses to embrace F/OSS artefacts, and create a self-enforcing quality enhancement system within the community.

In order to focus the study on solutions that would be acceptable by the community, further examination of the F/OSS development processes is required. As mentioned earlier, F/OSS projects are usually initiated by a developer who recognises a need and decides to develop an artefact to fulfil this need. In this case the developer is also the user of the artefact.

The current initiation process marginalises user involvement to a large degree, with the exception of some projects that were initially started as a response to user needs, mostly by commercial F/OSS entities. However, if the aim of the proposed solution is to signal quality to end users in order to increase their confidence in the artefact, user participation in the requirements determination is essential. There is no easy way to guarantee that the artefact will satisfy user requirements if those requirements have not been made explicit and are not clearly defined.

Upon determination of the artefact's requirements, the project initiator will become its maintainer and accept or reject any contributions made to the project. Maintaining a F/OSS project is a difficult and demanding task, especially with the increasing size of the project and the growing number of contributors. Ideally, the project maintainer is expected to check and validate every single contribution to decide whether it should be accepted and included in the final artefact, or whether it should be rejected because of poor quality, malicious intentions, or other reasons. This could be performed for small sized projects where the number of contributors and the load of contributions can be controlled by a single developer (the maintainer) within an acceptable time frame. When the project size increases, different approaches should be considered.

This issue has been dealt with by different projects using various techniques; some increased the number of the project maintainers while others considered a modular architecture where each sub-project could have its own maintainer (in addition to several other techniques). Essentially, the quality of contributions should be assured, as these directly affect the quality of the final artefact. This leads to the first question:

Figure 1. Coordinator, contributor, and end user roles in the F/OSS community

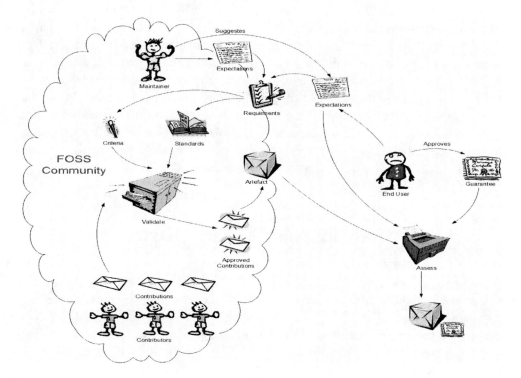

- How can the maintainer(s) of a F/OSS project accept contributions from community members in a timely manner while maintaining the required quality levels?

After the maintainer(s) incorporate the accepted contributions within the artefact, and decide that it has enough features to be released, the artefact could be made available to the user. The proposed solution should enable the maintainer(s) to assure the user that this artefact will satisfy the requirements set forth, so the user can assess the artefact against their own needs. This leads to the second question:

- How can the F/OSS community assure end-users that its artefacts will satisfy their requirements in order to encourage them to embrace these artefacts?

The rich picture in Figure 1 visually depicts the scenario under consideration. The figure demonstrates the various processes involved in the F/OSS community, and in particular the types of interactions that take place between a project maintainer, various contributors, and a non-developer end user. As illustrated, a project maintainer makes suggestions about possible capabilities that a given artefact should have as a set of requirements. End users interact with the description provided by the maintainer, and evaluate their expectations about these requirements and the eventual artefact which has yet to be developed. A set of contributors can now work towards these requirements and also evaluate the expectations proposed by end users by offering their own versions of the artefact to a maintainer. It is now the responsibility of the project maintainer to evaluate these contributions against the criteria set out as part of the requirements, and to validate these contributions. Once these contributions have been validated by a maintainer, the end user can then assess the developed artefact against their own expectations and provide feedback.

Modelling and F/OSS Metrics

In order to provide appropriate answers to the previous questions, which capture the essence of the quality issues in the F/OSS development process, modelling techniques were used to develop various conceptual models to illustrate the system that would provide such answers. These conceptual models were then analysed to provide better understanding of the problem. Recommendations for suggested appropriate courses of action were made based on these findings.

The developed models served as the basis for further analysis, which examined the existing development processes and activities in the F/OSS communities, how they currently provide for quality assurance, and what changes could be introduced to these practices in order to improve the quality of the resulting artefacts. This involved relating the derived processes to particular metrics that provide evidence-based quality assurances to end users. Such assurances should be such that an end user considers them as sufficient proof of the suitability of a particular F/OSS artefact for a specific set of requirements.

The problem at hand is characterised by complexity, involving multiple interacting elements and roles. The whole process is enacted by interacting human activities and aims to satisfy a particular need (the development of a specific artefact that fulfils a specific set of requirements). Existence of humans complicates the situation as every human actor has his or her own perception of the intended aims of the process, and his or her own distinct expectations. This results in the process being perceived differently by the different stakeholders, with sometimes contradicting viewpoints (Checkland, 1999)

These characteristics of the problem situation seamlessly fit within Peter Checkland's definition of a "purposeful" human activity system (HAS) (1999). Checkland claims that situations with human activity systems are highly unstructured. Building structure into the problematic situation itself is almost impossible, as the problem does not exist in the real world; it exists in the observer's perception of the real world, which is the main distinction between soft and hard systems. Therefore, any account of the problematic situation should be based on a specific perception (Weltanschauung or world view). This description perfectly applies to F/OSS, as the different perceptions of the various roles result in a distinct understanding of the problem. Checkland (1999) proposes the soft systems methodology (SSM) as "a general problem solving approach appropriate to human activity systems."

Choosing SSM as the methodology of choice to tackle the problem was based on the previously discussed features characterising the problematic situation. These features render the use of systems engineering techniques highly ineffective. Using quantitative systems methods was avoided because of the difficulty associated with the determination of measurable factors.

SSM suggests an intellectual construct called "root definition" (RD), which aims to provide a clear and unambiguous textual definition of the system to be modelled. It provides a way to capture the essence (root) of the purpose to be served (Wilson, 2001). Different root definitions could be developed to accommodate the different viewpoints or activities of the system.

Once the root definitions considered to be relevant were developed, they were used as the basis for building a conceptual model to accomplish what is defined in the root definitions. The conceptual model illustrates what the system should do to be the system described in the root definitions. It includes all the activities required and the logical links among them to achieve that goal.

We developed two root definitions to address the quality issues in F/OSS identified previously:

- **RD-1:** A system to provide guarantees acceptable by end users that the developed artefacts within the F/OSS community will meet the users' expectations by deriving appropriate mechanisms to assess these artefacts and produce such guarantees, while considering the resource constraints inherent in the F/OSS community.

- **RD-2:** A system to ensure that artefacts developed within the F/OSS community will meet determined quality criteria by deriving mechanisms to evaluate and approve submitted contributions to these artefacts against the acceptance criteria.

Based upon the formulated root definitions, a conceptual model that represents the system described in these root definitions was developed. The conceptual model illustrates any activities that are required to specify the system described by the root definitions, in addition to any logical relationships between these activities. The resulting conceptual model is reproduced in Figure 2.

The conceptual model provides valuable insights, as it illustrates the required processes that should be identified and implemented to achieve the purpose captured within the root definitions. Comparing the required processes and the situation in the real world revealed a set of possible actions that could be introduced to improve the problem situation. Identification of these processes helped to elicit requirements for the supporting information system, which in turn was utilised to support the implementation of the proposed actions for improvement. From Figure 2, the "C.A." arrow in some activities denotes that this activity should undertake control action to ensure the achievement of the system goals.

Each activity in the conceptual model was evaluated to determine whether it currently exists in a F/OSS process and how effectively it is performed. Such evaluation identifies missing activities that should be added and activities that could be improved. This is performed by mapping each activity in the conceptual model to the real world of the F/OSS community.

Figure 2. SSM conceptual model

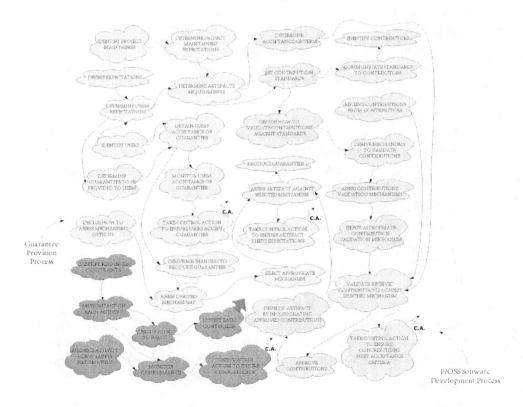

One weakness of SSM is that it only provides insight into what should be done to improve the problematic situation, and does not provide any guidance on how to build the required support system or implement the proposed solution. In order to describe the proposed solution in such a way that facilitates its implementation, different tools and techniques should be used in conjunction with the artefacts of SSM. Several studies have attempted to bridge this gap, including the work of Bustard, Zhonglin, and Wilkie (1999) and Wade and Hopkins (2002), who argue that SSM lacks the detailed information required to develop the intended system, and that the use of the Unified Modelling Language (UML) would compensate for this weakness and provide the level of detail required to facilitate the design and implementation of the desired system. Mingers proposed embedding structured methods into SSM (Mingers, 1992).

Two primary processes could be identified within the conceptual model, the boundaries of each process are illustrated in Figure 2:

1.   The F/OSS development process
2.   The guarantee provision process

Each process was identified by grouping appropriate activities relevant to the achievement of the process goals. Links between activities determine the logical progression of action and the required resources and expected results for the achievement of each activity.

To enable implementation of any proposed interventions that resulted from the SSM analysis, it was decided that UML should be used. This was undertaken in order to create models that describe the proposed solutions in sufficient detail to enable subsequent design and implementation of the supporting information system. Using UML's standardised notation assures the clarity of representation and ease of communicating the designed systems. The synergy of complementing SSM and UML will enhance the final outcome, and provide a logically defensible development route from high level system analysis down into specific system implementation details (Wade & Hopkins, 2002; Bustard et al., 1999).

The UML activity diagram for each process can be developed by mapping each activity within the process boundary in the conceptual model (Figure 2) to an activity in the UML diagram using its standardised notation. Flows in the UML diagram should match the dependencies represented in the conceptual model and the logical sequence of activities. The activity diagram also illustrates the actor who should perform each activity in addition to the resource requirements and outputs generated by each activity. Using UML modelling makes actors and outputs more

Figure 3. UML activity diagram of the guarantees' provision process

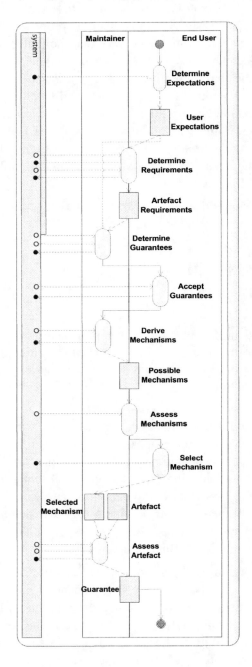

explicit and therefore provides the basis for the design of the supporting information system. Figure 3 shows the UML activity diagram developed for the guarantees' provision process.

User Participation

Analysis of the conceptual model indicates a significant lack of end user involvement in the current development processes within the F/OSS community. Unfortunately, the existing development mechanisms consider all users to be developers, and expect them to start their own development projects if they have any special needs or requirements. Lately, this situation has started to change slowly, with some communities and commercial organizations developing F/OSS software realising the need to get users' involvement in the development process in order to create software that will better satisfy users' needs. Ubuntu is building a GNU/Linux distribution based on interaction with users to determine their expectations and requirements (2005). The use of user-centred design techniques in RedHat Linux is another example (2005).

In order to improve the quality levels of the F/OSS artefacts, end user requirements and expectations should be considered from the early stages of the project development process. Quality may be perceived differently by different parties (Michlmayr, Hunt, & Probert, 2005), and our proposed approach suggests a systematic method to consider and integrate the different perspectives of developers and end users.

Upon determining the expectations of a maintainer and user, the maintainer will try to decide whether the user expectations are in line with his own. At this point, the maintainer may alter his own expectations in order to align them with those of the user, or if there are substantial differences between the two, the maintainer may not alter his expectations, and the user might look for another project that better matches his expectations.

After a balance is struck between the maintainer and user expectations, the maintainer will translate these expectations into detailed requirements. In the current processes, the maintainer is translating his own expectations into requirements, without considering those of the user. This process will be improved by considering the user expectations in accordance with the findings of the process analysis.

Requirements are descriptions of the "desired features of the proposed system." They may be functional (describing what the system should do) and non-functional (describing how a facility should be provided, or how well, or to what level of quality) (BSI, 1994). F/OSS projects usually lack an explicitly formulated requirements specification (Bartkowiak, 2004), because most of the projects depend upon the evolutionary development of requirements. Determining project requirements does not conflict with this concept, and the recommendation simply suggests specifying the translated expectations in clearly defined requirements that the project should meet. The initial formulation may capture only the essential required functionality, and could develop through the same current evolutionary mechanisms.

Contributions' Approval

The F/OSS development process relies on the contributions submitted by volunteer developers who might become interested in a particular project. Because of the volatility of such contributions, not much confidence can be placed in their quality. Therefore, the maintainer of each project should establish strict quality standards and acceptance criteria for the approval of submitted contributions. These standards and criteria should fulfil end user requirements and expectations, and be communicated effectively to all potential contributors.

The project maintainer is the final authority to decide on the overall direction of the project and approve contributions made to the project. Approved contributions to the project are incorporated in what is called the "official release." There is nothing that prohibits contributors from integrating their developments into the project even without the approval of the maintainer, but the maintainer cannot be held responsible for such releases. Usually when a conflict occurs about specific features in a project, a contributor or a group of contributors may initiate a new project.

Typically, the maintainer should examine every single contribution to ensure its compatibility with the main project and to confirm that it does not cause unintended side effects (either intentionally or accidentally) when incorporated into the artefact. However, this practice is very unrealistic. Due to the limited time and resources available to the maintainer, and with the increasing size of the project and its contributing community, it is almost impossible for the maintainer to check all the contributions submitted to the project. Therefore, project maintainers within the F/OSS community rely on informal trust mechanisms and on their own network of acquaintances and personal relations with contributors. These practices enable maintainers to place some degree of trust in contributions from people they know. They may also initially start dealing with a contributor's submissions sceptically, until this contributor establishes a reputation based on the quality of his submissions. Afterwards, the project maintainer could increase his or her trust in this contributor's submissions. Reputation may be transferable to other projects, although some projects might require specialist skills and therefore would render the past reputation of the contributor useless.

If the project maintainer intends to create high quality artefacts, the above mentioned mechanisms cannot be relied upon, as they lack a structured systematic basis, and are highly informal. Therefore, the maintainer should look for a way that would enable comprehensive testing and checking on the submitted contributions to be performed before these are included in the final artefact. By proposing automated validation tools, the acceptance process would become more effective. The project maintainer could evaluate and adopt appropriate validation mechanisms that are supported by automation tools. When a contribution to the project is submitted, it can be validated against the acceptance criteria and a decision made about whether it is useful to be incorporated into the final artefact. Validation results could be archived to enable tracking of any incorporated contributions.

Contribution standards and acceptance criteria cannot be determined beforehand or generalised for adoption in different projects, as they are highly dependent on the project and implementation technologies. They also depend on the desired quality level of the final artefact. Consequently, project maintainers should have the freedom in determining the required standards and the criteria they will use to define the acceptability of contributions.

Quality Assurance

In order to assure the end user that the developed artefact satisfies his or her requirements and expectations, the F/OSS community should provide sufficient guarantees to prove this compliance. These guarantees should be acceptable to end users, reproducible, and produced in an open and transparent manner. In consultation with end users, project maintainers must propose and select possible guarantees in addition to mechanisms that are required to produce these guarantees.

Automated test tools, where available, should be used to support the whole process (examples of such automated testing tools include SimpleTest for PHP (2005) and JUnit for Java (2005)), which should be kept as open and transparent as possible. This will enable any user to validate the claims of the maintainers and developers by conducting these tests using the same mechanisms that were used in producing them. Furthermore, users may customise the testing scenarios to match their own specific requirements.

Selection of appropriate testing techniques largely depends upon the implementation technology (programming language, operating system, etc.) and the availability of testing tools for that particular technology. The selected tool should be able to perform tests that would generate results that could be validated against the approved acceptance criteria.

Although different testing tools (both proprietary and open sourced) are available and could be used to perform software testing, it is recommended that project maintainers select F/OSS testing tools (e.g., those at www.opensourcetesting.org). The use of F/OSS testing tools increases the transparency and openness of the testing process, and complies with the ethical norms of the community, which in turn will facilitate the adoption of these testing techniques.

While software quality lacks a clearly defined notion, it is suggested that software has only to be as good as its users expects it to be (Bach, 1996). This means that for any application software to be (good enough) from its end user's perspective, it should satisfy the requirements upon which it was developed. Therefore, the existence of at least a requirement is inevitable for the evaluation of a software application's adherence to these requirements.

Unfortunately, current F/OSS development processes do not require the availability of any requirements or specifications, nor do they encourage user involvement in setting these requirements. However, the previously proposed actions suggest specific mechanisms to alleviate this problem and increase user participation in determining their expectations for F/OSS artefacts.

The final activity is the assessment of the developed artefact against the selected validation mechanisms in order to ensure the proper functionality of the artefact, and to produce guarantees that assure the end user of the conformance of the artefact to his or her expectations.

If the project maintainer validates the artefact against the selected mechanism and the artefact passes the proposed tests, he or she can release the artefact with greater confidence that it will attract better adoption by end users. Test results and the acceptance criteria should be bundled with the released artefact to enable the authentication of the guarantees by any potential user and to ensure accurate linking between different releases and their associated guarantees.

In case the artefact fails the selected tests, the project maintainer can trace the erroneous behaviour and attempt to correct it until the artefact can pass the required tests. This process will have great impact on the adoption levels of F/OSS artefacts within the end user community, as it increases their confidence that these artefacts will meet their expectations by providing them with evidence-based guarantees.

Although project maintainers could tweak their own project collaboration platforms to perform the proposed activities and processes, this will result in varying implementations of the same concepts, causing confusion to users. Building a central platform that provides these services in a standardised way and which project maintainers could use to host information about their projects would eliminate this confusion. This approach also offers the advantage of comparing different projects against each other.

Recommended Courses of Action

The recommended actions for enhancing the adoption levels of F/OSS artefact, by promoting evidence-based quality assurance mechanisms to be embedded into the current F/OSS development processes, can be summarised as follows:

- The F/OSS community should develop mechanisms to enable a higher level of user involvement in the development process.

- User requirements and expectations should be considered and clearly defined from the early stages of the F/OSS development process.

- Project maintainers should establish strict quality standards and acceptance criteria for the approval of submitted contributions.

- Project maintainers should evaluate and adopt appropriate validation mechanisms that facilitate the approval of submitted contributions.

- Automated validation tools should be utilised wherever possible to enable the acceptance of submitted contributions to the F/OSS project.

- Quality assurance guarantees should be identified in consultation with end users, and mechanisms to produce these guarantees in an open and transparent manner should be selected.

- Free and open source testing tools should be used wherever possible to provide the identified quality assurance guarantees.

- All testing and validation results should be bundled with the relevant F/OSS artefact to enable the authentication of the guarantees by any potential user, and to ensure accurate linking between different releases and their associated guarantees.

By involving users in the requirements and subsequent testing process, it is therefore possible to develop artefacts that are more closely aligned with user needs. We also believe that this approach will provide a greater level of confidence in quality and usability to end users of open source artefacts. There are no guarantees, however, that any resulting artefact will fully meet user needs and expectations. However, by making the user involved in identifying (and prioritising) requirements, it is possible to provide greater trust in the artefact that arises from such a process. The reputation of the artefact generated is therefore based on the reputation of contributors involved in the process and the transparency of the process itself.

Prototype

The main objective of the prototype is to demonstrate the use of the proposed supporting information system to support the quality assessment process. Considering the nature of the problematic situation, the distributed set of actors, and the openness of the F/OSS community, the proposed information system will be developed as a portal that provides the required functionality identified earlier.

The prototype was developed using the PHP scripting language (2005). PHP was chosen because of its particular suitability for Web development. PHP is also released under an open source license, which makes it a viable option for developing an application that is aimed towards the F/OSS community. The back end database uses MySQL (2005), an open source database management system that provides a high level of integration with PHP and other programming languages.

Prototype Implementation

Users or developers may use the portal interface to propose a new project and describe their expectations of the artefacts to be developed. If the proposal is made by a developer, he or she may also wish to initiate a project using the same interface; alternatively, the developer may browse the available proposals and initiate a project that interests him or her. Once the project is initiated, the developer becomes the maintainer of the project, and can manage its development using the project manager interface. The system also provides file upload facility to allow requirements descriptions to be added.

Using the project manager interface, the maintainer can add the requirements he or she defined based on the expectations found in the project proposal and request user feedback on these requirements. In order to facilitate requirements approval, the maintainer may define an approval threshold for each document. When the number of users who approved the requirements reaches the defined approval threshold, the document status automatically changes into "Approved." Once proposed, users can review the requirements and choose to approve it or not. The maintainer can monitor the status of any proposed requirements document from within the project manager. This feature satisfies the recommendation of requirements gathering from

Figure 4. Project requirements interface

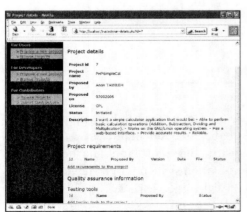 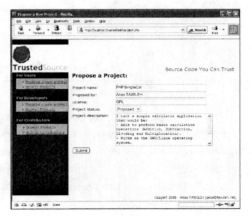

end users in the early stages in the development process. Figure 4 illustrates the project requirements interface.

The quality assurance section of the portal enables maintainers to suggest testing mechanisms and tools. These will be used for validating the quality of developed artefacts and produce appropriate guarantees to assure end users that the artefacts will meet their expectations. To enable the maximum flexibility in choosing a testing mechanism, the system does not impose any pre-defined structure on the organisation of proposals. It allows the maintainer to register any testing category (such as unit testing, integration testing, or performance testing) he or she might deem appropriate, and propose different testing tools to perform each one. Users are then given the opportunity to select the mechanism they find most appropriate for their purpose. The tool that achieves the highest number of approvals automatically becomes the selected tool. This functionality fulfils the requirements of the recommendation to identify and select mechanisms to produce quality assurance guarantees in consultation with end users.

After users approve the proposed testing tools and mechanisms, the maintainer can validate the developed artefacts against these mechanisms and provide the successful testing results as guarantees to end users. The system preserves the transparency of the whole guarantees' provision process, as the maintainer should upload the results of any test cases and the associated results as proofs of the appropriate behaviour of the artefact. Providing the test cases and the testing tools allows external users to validate the provided test results, and to reproduce the same tests and verify their outcomes. Figure 5 shows the quality assurance information interface.

The developed prototype also supports the contribution approval process by enabling the interaction and collaboration between the project maintainer and the potential or actual contributors through the contributions manager interface. Using this interface,

Figure 5. Quality assurance information interface

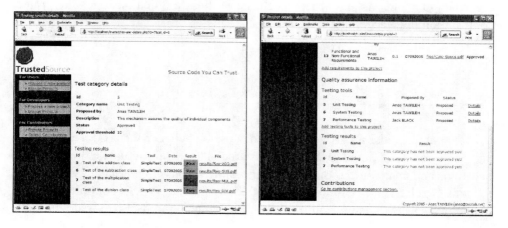

Figure 6. Project contribution interface

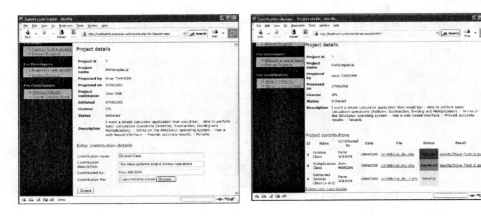

a maintainer can determine the required contribution standards based on the artefact's requirements. He or she can also define the acceptance criteria and communicate them to potential contributors using the system. Appropriate validation mechanisms to inspect contributed code for compliance can then be determined and entered into the system, to make them externally accessible and transparent.

Developers interested in a project may review the required contribution in light of standards and acceptance criteria, and decide to contribute to this project. The system facilitates this task by providing contributors with a simple interface to submit any contributions they develop. The status of submitted contributions remains *pending* until the project maintainer validates them against the acceptance criteria using the selected validation mechanism. The project maintainer makes his or her decision on the approval or rejection. He or she should include the testing results supporting his or her decision in order to explain the reasons of the decision to contributors and other developers. This will also facilitate the correction of any deviations from the contribution standards and acceptance criteria. Figure 6 demonstrates the project contribution interface.

Future Trends

This chapter provides a starting point for discussion between the F/OSS community and end users. Additional research in this area is necessary to tackle the identified questions and challenges proposed by this contribution. A significant question that remains is related to the introduction of change to the F/OSS community itself, including aspects that may contribute to the change in mindset required within the

F/OSS community in the way open source artefacts are developed, and the effects of the potential resistance to change within the community.

Further research may be conducted to investigate the possible ways of implementing the proposed guarantees into the final artefacts in digital forms, and the possibility of validating these guarantees without human intervention. Another possible area of research may include the integration of all the proposed testing mechanisms within a single Web-based user interface to allow verification of test results by end users. The proposed solution could also be enhanced by introducing quantitative evaluation and normalising the results between different tests and technologies to enable easier comparison of multiple artefacts. Improvements to the system may also include the implementation of a sophisticated feedback and rating system that collects user feedback, to demonstrate the actual utility of the artefact from an end user perspective.

Conclusion

Quality in the F/OSS community is a sensitive issue that has been out of focus for a long while. Although the F/OSS community proved to be highly effective in managing massively complex distributed development efforts, it did not provide proper consideration for the quality aspects of its development processes. This has inhibited the adoption of F/OSS outside the developers' communities, and created a highly sceptical perception about the viability of F/OSS for commercial and business environments. Most of the proposed initiatives to overcome this problem focused primarily on assessing the maturity levels of any F/OSS artefact by suggesting a formal analytical framework to gather and investigate information that is already available. Such initiatives did not suggest any interventions into the development process. Therefore, their results are highly subjective, and their contribution to the real improvement of F/OSS quality remains limited.

In order to realistically enhance the quality of the F/OSS artefacts, a completely different approach should be considered. Such an approach should focus on improving quality and propose possible courses of action that would enhance the development process and enable higher levels of user participation in order to boost the quality of developed artefacts-based on user requirements and expectations. Eventually, these new measures will result in increased user confidence in F/OSS, and, consequently, better adoption of its artefacts.

By introducing the proposed changes and implementing the prototype information system to support theses changes, it could be argued that F/OSS projects embracing this process would have a higher chance of acceptance by end users. The proposed approach will enable end users to communicate their requirements and expectations to members within the F/OSS community, thus facilitating development of artefacts that fulfil these requirements. The resulting increased level of acceptance are also likely to have an impact on subsequent F/OSS projects, thereby encouraging project coordinators to use the proposed platform and recommendations, or develop novel quality assurance and communication mechanisms. Either of these approaches will promote higher quality in F/OSS development and better adoption by end users. Hopefully, this will enable F/OSS to "cross the chasm."

In our approach, transparency and openness were emphasised during all stages of analysis in order to guarantee the production of a system that would be acceptable by the community as it adheres to its values of inclusiveness and participation. An open and transparent process will also increase users' confidence in the whole system and encourage them to trust its outcomes and participate in its development. Another implication is that by stimulating user-centric development processes, other aspects of quality that were overlooked by the current development practices (such as usability and ease of use) will be made possible, and the community will be guided towards aspects that are most important from an end user's point of view.

We believe that the proposed approach will lead to a significantly different way to think about quality issues in the F/OSS community–and one which is also likely to lead to a greater uptake of the artefacts produced.

References

Angelius, L. (2004). Who's guarding the guards? We are. *DevX.com*. Retrieved September 4, 2005, from http://www.devx.com/opensource/Article/20135

Bach, J. (1996). *The challenge of "good enough" software.* Retrieved September 4, 2005, from http://www.di.ufpe.br/~hermano/cursos/calcprog/good-enough-software.htm

Baker, M. (2005). *SimpleTest*. Retrieved September 4, 2005, from http://www.lastcraft.com/simple_test.php

Bartkowiak, I. (2004). *Quality assurance of open source projects.* Paper presented at Open Source Software Engineering. Retrieved July 20, 2006, from http://projects.mi.fu-berlin.de/w/pub/SE/SeminarOpenSource2004QA/oss_qa_en.pdf

British Standards Institute. (1994). *BS 7738-1:1994: Specification for information systems products using SSADM (Structured Systems Analysis and Design Method).* Implementation of SSADM version 4.

Business Readiness Rating. (2005). Retrieved September 4, 2005, from http://www.openbrr.org/

Checkland, P. (1999). *Systems thinking, systems practice*. West Sussex, UK: Wiley.

Duijnhouwer, F. W., & Widdows, C. (2003). Capgemini expert letter open source maturity model. *Capgemini*. Retrieved September 4, 2005, from http://www.seriouslyopen.org/nuke/html/modules/Downloads/htmldocs/osmm1.html

Glass, R. (2001). Is open source software more reliable? An elusive answer. *The Software Practitioner, 11*(6).

Golden, B. (2004). *Succeeding with open source*. Indiana, USA: Addison Wesley.

Jones, R. (2004). Open source is fertile ground for foul play. *DevX.com*. Retrieved September 4, 2005 from http://www.devx.com/opensource/Article/20111

Feller, J., & Fitzgerald, B. (2000). A framework analysis of the open source software development paradigm. In W. Orlikowski, P. Weill, S. Ang & H. Krcmar (Eds.). *Proceedings of the 21st Annual International Conference on Information Systems* 58-69, Brisbane, Queensland, Australia.

Fuggetta, A. (2003). Open source software: An evaluation. *Journal of Systems and Software, 66*(2003), 77-90.

Lerner, J., & Tirole, J. (2002). Some simple economics of open source. *Journal of Industrial Economics, 50*(2), 197-234.

Moore, G. A. (1999). *Crossing the chasm: Marketing and selling high-tech products to mainstream customers*. New York: HarperBusiness.

MySQL Web Site. (2005). Retrieved September 4, 2005, from http://www.mysql.com

Nichols, D. M., & Twidale, M. B. (2003, January). Usability and open source software. *First Monday, 8*(1). Retrieved July 20, 2006 from http://firstmonday.org/issues/issue8_1/nichols/

Object Management Group. (2005). Retrieved September 4, 2005 from http://www.omg.org/

Object Mentor Inc. JUnit. (n.d.). Retrieved September 4, 2005, from http://www.junit.org/index.htm

Open Source Maturity Model. (2003). Retrieved September 4, 2005, from http://www.seriouslyopen.org/

Opensourcetesting.org. (2005). Retrieved September 4, 2005, from http://opensourcetesting.org

O'Reilly, T. (1999, April). Lessons from open source software development. *Communications of the ACM, 42*(4), 33-37.

Ousterhout, J. (1999). Free software needs profit. *Communications of the ACM, 42*(4), 44-45.

PHP Web Site. (2005). Retrieved September 4, 2005, from http://www.php.net

Raymond, E. S. (2000). *The cathedral and the bazaar.* Retrieved September 4, 2005, from http://www.catb.org/~esr/writings/cathedral-bazaar/cathedral-bazaar/

RedHat ACS User-Centered Design. (2005). Retrieved September 4, 2005, from http://ccm.redhat.com/user-centered/

Scacchi, W. (2002). Understanding the requirements for developing open source software systems. *IEE Proceedings—Software, 149*(2), 24-39.

Stamelos, I., Angelis, L., Oikonomou, A., & Bleris, G. L. (2002). Code quality analysis in open source software development. *Information Systems Journal, 12*(1), 43-60.

Ubuntu Web Site. (2005). Retrieved September 4, 2005, from http://www.ubuntu.com

Wheeler, D. A. (2005). *Why open source software / free software (OSS/FS)? Look at the numbers!* Retrieved July 20, 2006, from http://www.dwheeler.com/oss_fs_why.html

Zhao, L., & Elbaum, S. (2003). Quality assurance under the open source development model. *The Journal of Systems and Software, 66*(1), 65-75.

Chapter VII

The Chase for OSS Quality:
The Meaning of Member Roles, Motivations, and Business Models

Benno Luthiger, ETH Zurich, Switzerland

Carola Jungwirth, University of Zurich, Switzerland

Abstract

This chapter explains why software users have good reasons to trust in the quality of OSS, even if they might have internalised the rule "If something has no price, it also has no value!" We present the idea that a system of incentives of both private programmers with their different motives to participate and companies paying their programmers for contributing to OSS, are responsible for the software quality—even if all programmers do not pursue a common purpose. The chapter delivers a conceptual framework from an economic perspective showing that every stakeholder can provide valuable input to the success of an open source project. Crowding out between contributors with different motivations does not necessarily exist even if companies with monetary intentions participate. Therefore, we assume OSS as an attractive forum for different interests that can seminally intertwine, while quality software is generated nearly as a by-product.

Introduction

Open source developers produce software—frequently of high quality—that is freely available for everyone. For computer users that are accustomed to purchasing outright the software they use, this might sound puzzling: Is it possible that something *free* is at the same time *good*? Therefore, such computer users, assuming a trade-off, might doubt the quality of open source software and hence refrain from using such software. Computer users that apply software in a business context, for example, may abstain from using open source software even if such software is available at no cost if they are not confident about the quality of such software. Using software in a business context means a heavy investment, even if the license fee of the software used is null because considerable TCO (total costs of ownership) exists. Integrating certain software in the business process leads to a lock-in situation in a manifold way (e.g., investments in human capital, system reorganisation, etc.). Thus, crucial for the future success of open source software is not the fact that there is no associated licensing fee, but the question of the quality of such software.

What can we learn about open source software if we look at developers motivation? Understanding the motivations of open source developers allows us to assess the importance software quality has for these software developers and to comprehend the conditions needed for that software quality to be realised. This insight might increase confidence in the quality and sustainability of open source software and, as a result, lower the barrier to use it.

Although examples of high quality open source software are well known, open source still has the connotative impression of being the play area of hobbyists. This impression is not necessarily far from the truth, insofar as most of the open source projects existing on open source platforms (e.g., SourceForge) are indeed the outcome of hobbyists, as various studies show (see Krishnamurthy, 2002; Weiss, 2005, etc.). However, the numbers in our FASD study[1] and in other studies (e.g., Lakhani & Wolf, 2003) indicate that professional developers who are compensated for their work do a significant share of open source software development. Indeed, the commitment for open source projects primarily occurs in their spare time; on the other hand, the share of open source projects developed within working time amounts to a considerable 42%. Professional open source projects might be underrepresented in previous studies, like our FASD study, because firms can afford their own project infrastructure and, therefore, are not dependent on the open source platforms we have addressed in our study.

This recognition shows that the motivations on the programmer level on the one side and the motivations on the firm level on the other side should be distinguished. The former concerns developers who commit themselves to contributing to open source projects. The latter concerns developers being motivated simply by the fact that they are advised to create open source software and they are paid to do so. In this case the interesting question is why their employers pay them to work on open source software. This should be a logical consequence of the business models for firms that build on sponsoring open source projects. However, their motivation should be discussed later. First, we analyse motivations on the programmers' level.

Analysing Motivations on the Programmers' Level

One way to resolve the puzzle concerning the open source phenomenon is to focus on the programmer as "prosumer" (see Toffler, 1980). A prosumer is a user who adapts and refines the software according to his or her needs. Von Hippel and von Krogh (2002) showed that such a point of view provides a substantial insight into the understanding of an open source development process.

And it also allows for maintaining the notion of rational actors. Prosumers do not act less rational then software firms investing in software. Both types of actors, the prosumer and the software firm, release the software under an open source license, if the benefits exceed the costs of such an action. Nevertheless, the prosumer's context differs fundamentally from a software firm's context. Software firms are in stiff competition against each other. In such a situation, uncompensated code releases imply high opportunity costs: By revealing the source code, software firms give away their business secret and, therefore, abandon a competitive advantage.

User-developers face a different situation. The interaction of prosumers is determined by only low rivalry conditions and, therefore, by low opportunity costs. Additionally, the direct costs of giving away the source code are rather small considering the Internet. However, small expenses on the cost side cannot explain prosumers' behavior. Even with small costs, it would be superior for rational actors to participate as free riders in the open source area and profit from others' work rather than to engage actively in the creation of such software. Therefore, there must be a selective incentive for contributors—a benefit that only persons who engage can reap. However, in low cost situations only a small selective advantage is needed and the free revealing of the source code is favourable for the user-developer. To understand the open source phenomenon on the programmers' level, we have to identify and quantify the possible selective advantages of open source developers:

A lot of work is done in this field and qualitative studies have identified a variety of motivations that can serve as selective incentives for open source developers.

Use

The simplest reason to engage in an open source project is that developers need a certain application. This is a straightforward implementation of the prosumer model: An actor has a problem that he or she could solve via suitable software. Therefore, they create that software or adapt existing open source software according to need. In the study of Lakhani and Wolf (2003), this motive was named most often. In our study we asked the respondents why they started their open source engagement. Running a cluster analysis, we could identify three distinct types of open source developers and one of them is clearly motivated by pragmatic reasons. The share of pragmatic developers in our sample amounted to about one-fourth.

Reputation and Signalling

In his essay, "Homesteading the Noosphere," part of the well-known trilogy "The cathedral and the bazaar" (2000), Raymond describes how norms and taboos affect the gain of reputation within an open source community. Raymond demonstrates that code forking and, above all, the removal of the contributors' names from the applications' credit files are strongly proscribed. The awareness of the open source community towards these norms makes sense, considering the gain of reputation. Without these norms, it would be hard to track which contribution comes from which person and this would hamper reputation building. Raymond concludes that these norms allow for a reputation game that is essential for the open source move-ment. It rewards the productive, creative, innovative contributors and thus founds the success of the open source movement.

Lerner and Tirole (2002) interpret the reputation game as signalling incentive. According to their view, the disclosure of the source code in combination with the specific norms governing the open source community as described by Raymond is quite attractive for developers for the following reason. It is very easy tracing the codes of developers back to them and assessing the quantity and quality of their contributions. A developer's status within the project depends on their performance and, therefore, reputation reflects skills, talent, engagement, and all other important characteristics from an employer's point of view. This coherence allows programmers to convert their reputations into cash, for example, by finding a better job offering or by better access to venture capital. One might ask why a firm does not assess the quality of the programmers' skills and talent by itself. However, this is difficult for a person who is not familiar with a special field. If a person's reputation is a valid indicator of his or her talent, this reputation can act as a signal in the sense described above: From the open source project and the person's status within this project, a potential employer can make valid conclusions of the person's talent. Signalling has the best effect in an area with great technical challenges, where the relevant community (e.g., the peer group) is technically experienced, able to distinguish between good and outstanding performance, and capable of esteeming performance and ability (Weber, 2000; Franck & Jungwirth, 2003). In the case of open source, these conditions are exceedingly accomplished. Hann et al. (2004) was able to empirically prove that the status achieved in open source projects under the Apache umbrella had significantly positive effects on the programmer's job income.

Community Identification

Persons perceive themselves not only or not always as independently acting individuals, but they also feel and define themselves as members of a specific group. Therefore, they behave according to the norms and standards of this group. Identification with a group and its goals can explain an individual's actions (Kollock & Smith, 1996). Hertel et al. (2003) examined empirically whether this phenomenon does play a role even in the open source area. In their study among Linux developers, they asked the programmers about various aspects of their activity and correlated this information with data about the developers' engagement. Indeed, using statistical methods, they could verify that a significant amount of the developers' engagement can be explained by their identification with the developer team.

This result has been affirmed by the study of Lakhani and Wolf (2003). In their hacker survey they examined how group identification affects the developer's time engagement. They observed a significant positive effect. Even in the FASD study, we identified a type of contributor motivated by the social context. About one-third of the respondents in the FASD sample belonged to this type (Franck, Jungwirth, & Luthiger, 2005).

Learning

Each social movement offers its participants the possibility to learn and to acquire special capabilities. This aspect is even more pronounced for the engagement in an open source project. Open source projects, furnished with the appeal of programming at the edge of technological innovation, promise to offer extraordinary learning opportunities. In addition, the peer review system specific for the open source area provides timely feedback (e.g., identification of software bugs or suggestions for code improvements) that increases the contributor's learning effect. The desire to improve one's skills as a software developer appears in various studies (see Ghosh et al., 2002; Lakhani & Wolf, 2003; Hars & Ou, 2001, e.g.).

Altruism

Programmers sometimes engage in an open source project with motivations that can be described as altruistic. They contribute, for example, because they use open source software and, thus, feel the obligation to reciprocate. In other cases, they contribute with the intention to aid other people, for example, in developing countries because freely available software helps to bridge the digital divide. In this case, the utility of the open source programmer is increasing with the other's benefit (pure altruism) whereas in the first case the programmer might feel a "warm glow," indicating impure altruism, by doing the right thing (Haruvy et al., 2003). In any case, such contributions can be interpreted as donations. The logic of this interpretation is that making the contribution or not does not have any traceable consequences for programmers themselves. On the other side, the open source project profits from such contributions.

According to the study of Lakhani and Wolf (2003), about one-third of the respondents indicate such motives as relevant for their engagement. Interestingly, Lakhani and Wolf could show via cluster analysis that the ideologically-based argument that software shall be free is close to the reciprocity motive. That would mean that such ideological motives could have an altruistic connotation, too.

Fun

Most of the persons developing software perceive this activity as exceedingly fulfilling: "Programming [then] is fun because it gratifies creative longings built deep within us and delights sensibilities we have in common with all men" (Brooks, 1995, p. 8). Torvalds (Ghosh, 1998, p. 9) said as follows: "[M]ost of the good programmers do programming not because they expect to get paid or get adulation, but because it is fun to program. [...] The first consideration for anybody should really be whether you'd like to do it even if you got nothing at all back." Thus, developing software can have an immediate benefit for the programmers that can be named *homo ludens payoff* in accordance with Huizinga (2001).

Open source developers program in their spare time because they consume "fun" with this activity and, therefore, open source software is a by-product of this activity. Indeed, several empirical studies corroborated the importance of fun as motivation to engage in open source projects. As an example, the study by Lakhani and Wolf (2003) showed that 73% of the open source developers experience flow while programming. Although this explanation sounds reasonable, it's not enough to justify the existence of open source software. The fact that programming can be a fun activity independent of compensation explains only the existence of software developers in general. However, we have to take into consideration that one can earn money by developing software. If we are dealing with rational software developers, the possibility to earn money is without doubt an additional utility. Consequently, we do not expect anyone, at least no rational actor, to program in his or her spare time anymore, because having fun *and* earning money is *mutatis mutandis* better then only having fun. Therefore, if we want to explain the existence of open source software by the fun motive, we have to postulate that having fun while programming is somewhat substitutive to earning money with software development. The open source development has to offer better opportunities to enjoy programming then working in the commercial software area.

In the FASD study, we focused exclusively on the fun motive. The aim of the study was to quantify the importance of fun as motive to engage in open source projects. In addition, we also tried to verify the hypothesis that programming provides more fun in an open source context than under commercial conditions (Luthiger, 2006).

To quantify the importance of fun, we looked at the variation in the open source developers' engagement and inquired how much of this variation can be explained by the variation of fun the developers enjoy while programming. Thus, the task to quantify the importance of fun becomes an exercise in regression analysis. To master this task, we developed a simple model combining the open source developer's engagement with the fun he or she enjoys and the amount of spare time he or she has. We used a production function whose input factors, in our case the fun and spare time, have diminishing marginal effects on the output factor, the programmer's engagement. This can be achieved with quadratic terms having negative signs:

$$E = c + a_1 * F - a_2 * F^2 + b_1 * T - b_2 * T^2$$

where

E: voluntary, unpaid engagement

F: fun

T: spare time

a1, a2, b1, b2 > 0

To operationalise the fun developers generally have while programming, we used the flow construct introduced by Csikszentmihalyi. For the developers' engagement we used two measures: First we determined the amount of hours the developers spend in their spare time for open source. Second, we asked for their willingness for future engagement in open source projects.

The first regression analysis yielded the result that flow contributes significantly only linearly to the amount of time the developer spends for open source, whereas concerning the availability of spare time, both terms contribute significantly. This means that the joy of programming does not wear off: Each additional unit of fun is transferred linearly into additional commitment. Another result is that the amount of time spent is controlled about ten times more by the available spare time than by the joy of programming. Obviously, the limiting factor concerning the amount of time spent is not the fun experienced while programming but the available time of the programmers. With this model, we're able to explain 33% of the variance in the amount of time the open source developers contribute.

If we look at the determinants considering the willingness to engage for open source in the future, we get another interesting result. Again, flow contributes to the model significantly only with the linear term. This time, however, the available spare time does not contribute to any of the terms significantly. We conclude that when open source developers evaluate their willingness for future engagement, they take into consideration only how much they enjoy programming and neglect completely whether they will have the time needed for future engagement. With this model, we're able to explain 27% of the programmer's open source engagement.

The assumption that programming in an open source project provides more fun than doing this activity under commercial conditions can be tested by comparing the answers of our survey addressing open source developers with those of software developers working in Swiss software firms. We found out that indeed the open source developers enjoyed significantly more fun while programming than commercial software developers. To allow for a systematic bias coming from the comparison of two different samples, we looked for a possibility to compare the experience of flow within the sample of open source developers. Based on the answers about how much of their time they are paid for programming, we have been able to identify two sub samples. We named the first "professionals" because this sample consisted of open source developers that are paid for more than 90%t of their working time in open source projects, whereas the second sample, the "hackers," are paid for less than 10% of their time spent for open source projects. Therefore, our "hackers" stand for the classical open source developers, the "hobbyists," spending their spare time to develop open source software. The comparison of these two samples yielded again that the "hackers" experienced significantly more fun than the "professionals." This result confirms our conjecture that fun is an important driver for the creation of open source software.

Analysing Motivations on the Firm's Level

Because of their own infrastructure, open source activities of firms are underrepresented in the FASD study as well as in other studies. Nevertheless, evidence exists that paid software developers create a significant part of open source software. What incentives do employers have to pay software developers who create a product that is given away for free?

There are two main reasons for an employer to do so: The first reason is that the firm needs a certain software solution for its own use. By opening the source code application or joining an open source project, the firm can lower costs and spread risks. The second reason is that the firm has a business model that builds on open source software.

Use Value

Software developed in-house that has use value for the company and that does not represent any core competency of the company should not cause any losses if the source code is opened. In fact, Raymond (1999) identified two cases where a company can win by doing so.

If a firm operates a Web platform, for example, for selling low margin services or products, it usually needs a Web server, some kind of content management system, and a database. Nowadays, no firm would even think to develop such applications or to pay for the development of such applications. Instead, the firm takes one of those excellent applications released under an open source license and adjusts it to its specific needs. But then, to a certain extent the firm becomes dependent on the continuing existence of this software. Because of its investments in the software, the firm is strongly interested in its survival. In such a situation, it could be reasonable to improve the application's attractiveness and, hence, its user base, by code donations that expand the software's stability or functionality, for example.

Besides this possibility to lower costs, there is also the prospect of risk spreading by open sourcing code. Raymond (1999) exemplified this option by a story from a firm that developed in-house a special print spooling application. After putting the application into operation, the firm decided to release this print spooler under an open source license. The firm's ulterior motive was to stimulate an improvement process for the software. By releasing the code, noticeable problems become evident, other applications are found, and missing features are developed. In essence, the community should be interested in the software and further develop it. Without this move, the firm would have run the risk of letting the application become unmaintained, that is, to let it gradually fall out of sync with technological progress. By releasing the application, the firm could spread the application's maintenance over various independent contributors, thus minimizing the risk that the application goes out of date. If different independent stakeholders are interested in the survival of a software application, the probability grows that the software is kept up to the status quo of the technological progress, whereas none of the different stakeholders can privately appropriate the code.

Business Models

Open source software that is freely available poses a serious threat to the sales value of software. Nevertheless, good reasons exist why it can be worthwhile for firms paying developers to create free software. So-called "business models" describe different situations where firms profit from investing in open source projects. The following discussion bases heavily on the considerations made by Raymond (1999); Hecker (1999); Leiteritz (2004); O'Mahony et al. (2005); Weber (2004); Kalla (2005):

- **Open source application provider:** Such application providers create software that they distribute under the terms of an open source license. An "Open source application provider" is a generic term, in which various variations of this business model exist. Most of them succeed based on the existence of a complementary product or service (e.g., "Loss Leader," "Sell it, Free it," "Widget Frosting," "Service enabler"). The basic pattern to generate profits is that by giving away the software for free, the company enlarges the application's user base thus increasing the market for the complementary product.

- **Loss leader:** In the "Loss Leader" model, an application is given away as open source software to improve the company's position in the software market. According to Hecker, the open source product could increase the sales of the complementary software product "by helping build the overall vendor brand and reputation, by making the traditional products more functional and useful (in essence adding value to them), by increasing the overall base of developers and users familiar with and loyal to the vendor's total product line" (Hecker, 1999). Netscape's open source strategy with the Netscape/Mozilla Web browser is an example of this business model.

- **Sell it, free it:** In this variation, the application is sold (i.e., distributed with a commercial license like any commercial product) when it is ready for release. In a later part of the application's life cycle, for example, if the software company has developed a new version of the application, the application's source code (i.e., the older version) is opened. In such a model, the customers buying the software are paying a premium for the value of using the application earlier rather than later. This makes sense when the application introduces a functionality that is novel in the software market. After opening the code, the freed version can act as a "Loss Leader" for the application's new version. The later versions of the application can be built on the code of the earlier open source versions. To make this possible, the open source license chosen has to be liberal, that is, it has to allow that derived work can be distributed under a commercial license.

- **Dual licensing:** This is a business model similar to "Sell it, Free it" in so far as the application is available both under a commercial and an open source license. In this model, however, the application is simultaneously distributed under both license schemes. The two versions of the software address different target groups. The free version is intended for users that get familiar with the software by installing and using it and thus preparing the market for it. The open source license chosen for the application's free version has to be restrictive (e.g., General Public License GPL). Thus, software companies that want to integrate the software into their own applications need the software version with the commercial license. The code base of the two versions is the same but the version with the commercial license delivers additional support or product guarantees (in addition to the right to integrate the software). The well-known MySQL database is a good example of this business model.

- **Widget frosting:** In this model, the complementary product is hardware. For example, a printer manufacturer might release the drivers for their printer under an open source license, thus gaining a larger developer pool and better driver software. In the end, this improves the printers' acceptance and, therefore, a better market position for the printer manufacturer. In a way, Linux represents the open source software to sell Linux computers, that is, computers preconfigured with Linux, specially designed for an optimal support of this operating system. In fact, some companies do exactly this, and consequently, sponsor the development of the Linux software.

- **Service enabler:** In this business model, the complementary product is neither software nor hardware but a service that generates the company's revenue stream. The company sells a service online and needs software so that users can access the service. If the community enhances the client software and makes it more user friendly or ports it to new platforms, the market of this service will be expanded.

- **Standard creation:** If a company wants to create a technical standard it can safely use as a foundation to build its proprietary applications, open source can play a crucial role. A company sponsoring a standard faces a serious problem: In order for the initial work to evolve into a standard, it has to be taken up by other companies. How can other companies be convinced to adopt this standard and to contribute to it? The company that sponsored the code that builds the new standard can level the playing field for potential competitors by open sourcing this code. This works because within an open source environment, the competitors do not have to fear that the initiator can exploit hidden features creating software that is superior to those created by the competitors using the same standard. Moreover, by making the code open source, the initiator signals that contributors can participate in the negotiation about the future evolution of the standard, thus providing incentives for other companies to join it.

This strategy seems only possible for big companies having a long-term policy and the perseverance to pursue it. On the one hand, standard building needs several years of high investment without any return. On the other hand, to appropriate the gains of an established standard to an extent that exceeds the primary investments, the company needs a full portfolio of products and services that can be related to the new standard. An example of this model from practise is IBM's sponsoring of the Eclipse open source project.

The business models above described situations where the companies pay software developers to create open source software that constitutes the basic part of their business model. In other business models the company does not create software but profits as a free rider from open source software and the open source movement. However, by selling their services, they popularise open source software in many ways. Therefore, companies implementing such business models are rather symbiotically related to open source and, thus, well accepted in the open source movement:

- **Support sellers:** The principle of this business model is to sell support for users of open source software. There are two versions known for this business model: Distributors combine different open source applications to assorted and tested distributions (i.e., media and hard copy documentation) that can be sold. The famous company Red Hat pursues this business model very successfully. In another variant of this business model, companies sell technical support for users of open source software. This covers teaching, counselling, system integration, and system tuning, and so forth. The "Support sellers" business model is subject to risk for two reasons. First, the entry barriers of competitors in that market are very low; therefore, stiff competition drives the prices down. Second, the more open source software becomes user friendly, stable, and well documented, the less users of open source software have the need to buy support for such software.

- **Mediators:** The strategy of an open source mediator is to operate a hardware and software (e.g., collaborative tools) platform where open source projects can be hosted. Gains can be obtained by selling advertising space (banners). The "Mediator" model is characterised by a "winner take all" effect. The more successful a platform is in terms of the amount of participants, the more attractive it is for new participants. Having selected a certain platform, the developer's willingness to change to another platform is very small, especially if the other mediator hosts lesser projects and is frequented by lesser users. Furthermore, a challenging mediator has few chances to attack the leading platform because no battle on prices is possible since the services are free of charge anyway. Therefore, the market entry barriers for other mediators are rather high. Additionally, the costs of operation of a full-fledged open source platform hosting thousands of projects and frequented by thousands of users seven days a week are high, whereas the opportunities to create a revenue stream are limited. It is questionable whether the cash receipts from selling banners on the Web pages exceed the operation costs. The best-known and greatest provider of such a mediator service is SourceForge.

- **Accessorizing:** Companies pursuing this business model sell accessories associated with and supportive of open source software. T-shirts printed with the name and logo of a famous open source project or a professionally edited and produced documentation of open source software are examples of these products. O'Reilly pursues this business model and is well known for their various books on open source software.

In previous years the research community studying the open source phenomenon made remarkable advances. The research yielded interesting results in topics as the open source developer's motivations, the management of open source projects and the coordination of contributors, the importance of government funding, the consequence of software patents, and others. However, the relationship between open source software and the commercial software area remained largely unexplored[2]. Thus, there are some interesting research questions to answer:

Under which circumstances and in which situations are open source business models a viable option? What chances of survival have companies pursuing a specific open source business model? Are certain open source business models dependant on the type of software the open source project is creating (e.g., server, client, database, CMS, etc.)? What effects does the commitment of a business partner (with an openly declared business model) have on the quality of the software produced, on the release frequency, on the developer community, and so forth?

Analysing the Interplay of Different Motivations

Collective action problems describe situations in which everyone in a given group is confronted with a certain set of choices. If every member of the group chooses rationally in the economic sense, the outcome will be worse than if all members are willing to choose another, individually suboptimal alternative. Open source projects face collective action problems on different levels. The software produced is a public good, and therefore, the project has to deal with free riders in an *n-person prisoner's dilemma* situation: if the project succeeds and is able to deliver the software, everybody benefits. However, everybody can improve his or her situation by not collaborating. Thus, if everybody acts rationally, the project will not succeed and, hence, the software will not be produced. As we mentioned above, this first order social dilemma can be overcome by selective incentives. However, the problem to coordinate the different contributors remains. This is a second order social dilemma because coordinating the contributors and reinforcing the social norms guiding the collaborative work is a public good, too.

According to Elster (1989), social norms play an important role in overcoming the social dilemmas and promoting collective action. Concerning the possible attitudes toward social norms, Elster identified five different positions: (1) *rational, egoistic* persons are guided by the dominant strategy of non-cooperation; (2) *"Everyday Kantians,"* by following Kant's "Categorical Imperative," are guided by a norm that Weber named "ethics of conviction" (Gesinnungsethik), thus cooperating by all means; (3) *utilitarists* cooperate conditionally, if their contribution increases the average utility; (4) *elite-cooperators* contribute in an early phase of the project when there are few participants, whereas *mass-cooperators* contribute only after many other persons decided to cooperate. Both types share the common attribute that they have a private benefit not only from the results of the cooperation, but from the act of cooperation too; and (5) persons motivated by *fairness norms* contribute as soon as the general level of cooperation exceeds a certain threshold.

With such a set of motivational types, it is possible to explain the dynamic of collective action. Everyday Kantians act as catalyst for cooperation. Persons motivated by fairness norms can amplify this process. The number of utilitarists and persons motivated by fairness norms move inversely: The more persons contribute, the less effect there is from a utilitarist's contribution. Hence, their number is decreasing. At the same moment, an increasing number of persons motivated by fairness norms are induced to cooperate. The same happens for elite- and mass-contributors. Such a dynamic model establishes that persons who are differently motivated should be found in different phases of the open source project.

We can roughly distinguish three different project phases: project initialisation, development to stability where community building also happens, and maintenance phase with stable releases. Concerning the project roles, the following differentiation is useful: the *project leader* (also known as "benevolent dictator") is usually the project founder and has the ultimate responsibility. Among others, the project founder chooses the type of open source license for the project. Thus, the project leader normally has the right to relicense the open source project. The *committers* are contributing to the project's code base on a regular basis, the *lead users* are actively using the application, sometimes contributing bug fixes, adaptations, feature wishes, and the like, and *silent users* that use only stable releases, silently quitting as soon as the application does not meet their requirements any longer. Even though silent users do not contribute any code or information to the project, they might be important because of positive network externalities they create.

Using the motivational types we explained in the first section, we can depict the dynamics in an open source project as follows. Project initialisation is done by the project founder (we do not consider open source projects fully sponsored by companies here). The project leader may be of "everyday Kantian" type and moved by altruism. Stallman's founding of the GNU project may be interpreted in this light[3]. Project founders might also be moved by fun or by the need of the functionality. An example of this type is Torvalds and his Linux project. In this case, it's rather egoism then altruism guiding such project leaders.

In the project's next phase, the application's core functionality is created. In this phase of the project its core community has to be built while different types of committers join the project. They may be utilitarists as well as elite-contributors. Utilitarists contribute because they are interested in the result and their engagement helps the project. As elite-contributors they need a selective advantage from cooperation. This might be the fun they enjoy while developing for the project or the learning effect from contributing. At this stage, the project leader's attitude has to change gradually, at least if he or she has been moved by the fun motive originally. In order for community building to occur, he or she has to offer a credible project vision and challenging tasks for the developers joining the project. The more the project establishes, the more people join moved by fairness norms. Such persons are important to build community culture and identity. They also do the more tedious work essential to reach project stability, for example, project documentation, usability tests, quality and release management, and so on.

The "break-even" point referred to by project stability is the stage where the project has gathered enough momentum to attract reputation-motivated contributors. At this point, the prospects are good and the project will provide valid signals to outsiders in the near future. Concerning the social norms, such contributors are rather mass-contributors. At the same time, however, they need to be rather competitive, for that they can capture a position within the project that allows them to stand out from the other persons involved. It is well known from famous open source projects (Linux of FreeBSD, e.g.) that the entry barriers for new contributors are very high: Project leaders only accept codes that are unobjectionable and of outstanding quality. Thus, such projects indeed provide credible signals for the outsiders and, therefore, are attractive for programmers who play the reputation game.

The ultimate proof that an open source project is both stable and successful is its inclusion into a distribution. A distributor selects an open source application only if it adds value to his distribution on the one hand and if it is easy to install on the other. The first condition implies that the distributor has enough clues that silent users demand this application. The latter condition means that the project concerns not only about coding and architecture, but about documentation and packaging, too. However, being neatly packaged and included in a distribution for the delight of silent users bears the danger of stagnation for the project. Because of that reason, the project needs lead users so that it can evolve even in its stable form. Whereas silent users only work with the stable releases of an application, lead users download and install release candidates. Thus, they are acting as beta testers and provide helpful feedback to the project if they find bugs or deficiencies.

Lead users are elite-cooperators. They have fun using the newest version of a slick tool long before others; at the same time they learn and build up valuable knowledge about the application, its evolution, and hidden goodies and limitations. An interesting point is that whereas egoistic, rational persons do not participate in usual collective action providing a public good, even such people may contribute to open source projects, as long as they can consume enough fun, for example.

Which institutional arrangements are needed to foster such a dynamic of an open source project?

As described above, the first phase of an open source project can be explained either by altruistic or egoistic programmers. The first donate their work and time for a good aim. Even if the motives behind their donations are unidentified and possibly assessed as irrational, Franck and Jungwirth (2003) argued that donators choose a beneficiary who is "worth" the donation, for example, by credibly committing himself or herself to the nondistribution constraint. Thus, while pursuing their targets, donators act rationally. In the context of an open source project, an altruistic founder who contributes the initial code base to the public wants that the code donated will be free and not privately appropriated by commercial software companies. Thus, he or she wishes that nobody can turn donations into private profits. The institutional safeguard for this is the Copyleft clause of a restrictive open source license. The Copyleft not only ensures that the code donated will be free, but that additional work building on this code has to be free, too. Therefore, a restrictive licensing scheme efficiently prevents any attempt to commercialise on any such code.

A fun-seeking project founder, on the other hand, does not bother much about licensing. Indeed, having developed the initial code base, he or she has already consumed the *homo ludens payoff* and has, therefore, little reason to release the code at all. Thus, such project owners need additional incentives to do that. This might be that releasing code is cheap because of the Internet and because of platforms offering their hosting services for free. Another reason might be the pragmatic motive that the project needs a community when it reaches maturity. Considerations about the project leader's reputation might be an additional motivation to release the code. In all cases, the existence of both mediums, an active community that can be addressed and the Internet for communication, play a crucial role.

Elite-cooperators and utilitarists joining the project now may be moved by the fun motive. They have good chances to get satisfied, because in its early stage, the project might provide the most challenging tasks and there are only few developers competing for such tasks. To be attractive for these developers, the project's entry barriers must be low. A developer who scans through project descriptions looking for a nice challenge does not want to wait days until he gets his or her contributor's access. On the other hand, the environment has to be "forgiving." If one developer delivers a code piece that breaks the application's functioning in another module, this should not break down the whole project. Instead, it must be easy to roll back the code state and go on again from there. A forgiving and responsive environment can stimulate a vibrant developer community and lead to considerable results within a short time (see Broadwell, 2005).

In the later phase of the development stage, contributors enter the project who might be motivated rather by fairness norms than by fun. Just as for altruistic project leaders, for such contributors the license type is of importance. They donate their time and creativity to produce code, documentation, and so on and they wish that these donations should not be appropriated privately. Thus, a license having a copyleft clause provides the right incentives for such contributors.

Reputation-motivated contributors entering the project in its stable stage might have an ambivalent attitude towards the license of the open source project. On the one hand, the chosen license has to guarantee the visibility of their contributions. On the other hand, a too restrictive license might impede the application's diffusion, thus reducing both the project's and the contributors' reputations. In the end, they might prefer a liberal license as long as the project owner can credibly assure that he or she never re-licenses the project to a closed scheme. At last, it is rather the existence of lead users than a license issue that determines whether reputation-motivated contributors join or not. The lead users' feedback drives the project to a considerable amount, whereas a project without lead users will stagnate and decline within a short time. Lead users on the other side are attracted by new features. Therefore, reputation-motivated contributors and lead users have a reciprocal relationship: Reputation-motivated contributors implement the new features of an application, which the lead users demand, whereas the latter provide the feedback and stimulate activity. Therefore, lead users need a low-cost access to the source code or the application's installers as well as a credible signal that the open source project in its actual form will persist. Project failure or closing its source code will devaluate the lead-users' knowledge, so that uncertainty about the project's future will detract lead users from the project. Thus, a project whose code is owned by only one person is rather unattractive, whereas a project with broad code ownership or a code being held by a foundation is more attractive for lead users.

To conclude, as long as an open source project succeeds in accomplishing heterogeneous needs, it can attract the differently motivated contributors building a vibrant community required to make the project successful. To make this possible, both the project's license and the infrastructure are issues. For altruistic contributors, the copyleft clause is a prerequisite to attract them in the initial and the project's building phases. For fun-seeking contributors, the joy of programming depends on the infrastructure. Programming is usually more enjoyable, thereby enhancing the fun factor, if the developer can focus on coding without being distracted by organisational issues. To accomplish this positive environment where the developer can concentrate on what he or she loves to do most, the infrastructure has to be both highly available and responsive.

The remaining interesting question is, What happens to this arrangement when companies pursuing open source business models enter the scene?

We can distinguish between one scenario where the company donates the idea and vision of the application as well as its initial code base and a second scenario where the company joins an already existing project and establishs a business around the application.

In the first scenario, the company will not succeed in building up a community around the application until the project reaches stability. The project is not attractive for altruistic contributors because, even if the chosen license is restrictive, the company cannot credibly promise to let the source code open. On the other hand, the company cannot accept any code contribution from the outside without commanding the right to re-license the code. However, the company-sponsored project is of less attractiveness for the fun-motivated contributors, too. This is because such a project is driven by the in-house software developers and, thus, cannot offer challenging tasks to outside contributors. Nevertheless, having reached stability the company could try to build up a community by attracting both reputation-seekers and lead users. As we explained above, reputation-seekers might prefer projects released under a liberal license. However, implementing a dual licensing scheme might be a viable alternative. The commercial license makes the application fit for commercial use whereas the open source license with a copyleft clause guarantees the continuous openness of the source code, thus making the contributions visible. The same consideration holds for lead users.

Another strategy for companies planning to build up an open source community in the project's stable stage could be to hand over the code ownership to a foundation. Even this ensures the source code's continuous openness, thus making the project attractive for rent-seekers and lead users.

A company will join an already existing open source project most probably in its stable stage when the project has already attracted a community. For playing a role within the project, the company has to be accepted by the community and also be very careful not to scare away the community members. Such a strategy can be successful only if the company is very open and transparent about its intentions concerning the project participation. The company has to communicate clearly how it wants to earn money by promoting the project and at the same time why the openness of the code is vital for the company's business model. In addition, the company has to be careful that the project stays attractive for fun seeking contributors. This can be achieved if the company does not "in-house" the application's further development but lets the community develop and implement the application's enhancements.

The Pursuit of Software Quality

Software quality can be considered as consisting of code quality (e.g., testability, simplicity, readability, self-descriptiveness) and software usability (e.g., ease of learning, efficiency of use, error frequency and severity, subjective satisfaction). Stamelos et al. (2002) showed that code quality is acceptable for the Linux applications they tested. Nichols and Twidale (2002), however, assumed that open source projects suffer usability problems. How can the open source software development model achieve code quality and how could it improve on usability?

Pragmatic contributors want to see the improvements they added in an operational state and, thus, don't bother much about code quality. Fun seekers may enjoy "elegant," manageable code that is both simple and readable. However, the quality of their contributions depends largely on their ability and experience. If they are unskilled, they may experience fun producing "ugly," unwieldy code as well. Therefore, it's mainly the reputation-motivated contributors that drive open source projects' code quality because their reputation builds heavily on the quality of their contributions. In addition, their intensive interaction with lead users that act as beta testers enables them especially well for the pursuit of code quality.

Why is this same reasoning not true for usability? Lead users are computer experts no less then the contributors to the open source project. They don't need elaborate user-interfaces to fully exploit the functionality offered by the software. Therefore, they can't feed back usability issues to the project. Those who are best suited to evaluate the application's usability are the silent users. However, as they are silent, they don't give any feedback. Is there a way to capture the silent user's experience? Here's the place where companies entering the open source area while pursuing a business model can add value to the open source project. At least if they have a history in the software business, they are in connection with the application's potential and silent users and, therefore, can act as a proxy for them and provide valuable feedback concerning usability issues.

These reflections show that every stakeholder can provide valuable input to the success of an open source project. Crowding out between contributors with different motivations does not necessarily exist even if companies with monetary intentions participate. Therefore, we assume that open source development is not a temporary but rather a stable phenomenon because its particular production context allows every participant to put in and to put out as much (or as little) as he or she wants. Therefore, it is an attractive forum for different interests that can seminally intertwine, while quality software is generated nearly as a by-product.

References

Bonacorsi, A., & Rossi, C. (2005). Intrinsic motivations and profit-oriented firms in open source software. Do firms practice what they preach? In M. Scotto & G. Succi (Eds.), *Proceedings of the 1ˢᵗ International Conference on Open Source Systems*, Genova (pp. 241-245).

Broadwell, G. (2005). *-Ofun*. Retrieved May 4, 2007, from http://www.oreillynet. com/onlamp/blog/2005/10/ofun.html

Dahlander, L. (2004). *Appropriating returns from open innovation processes: A multiple case study of small firms in open source software*. Retrieved April 10, 2006, from http://opensource.mit.edu/papers/dahlander.pdf

Csikszentmihalyi, M. (1975). *Beyond boredom and anxiety*. San Francisco: Jossey-Bass.

Csikszentmihalyi, M., & Csikszentmihalyi, I. S. (1988). *Optimal experience: Psychological studies of flow in consciousness*. Cambridge, UK: Cambridge University Press.

Elster, J. (1989). *The cement of society: A study of social order*. Cambridge, UK: Cambridge University Press.

Franck, E., & Jungwirth, C. (2003). Reconciling rent-seekers and donators. *Journal of Management and Governance, 7*, 401-421.

Franck, E., Jungwirth, C., & Luthiger, B. (2005). *Motivation und Engagement beim OSS-Programmieren—Eine empirische Analyse*. Retrieved July 18, 2005, from http://www.isu.unizh.ch/fuehrung/Dokumente/WorkingPaper/36full.pdf

Gabriel, R. P., & Goldman, R. (2002). *Open source: Beyond the fairytales*. Retrieved October 4, 2003, from http://opensource.mit.edu/papers/gabrielgoldman.pdf

Ghosh, R. A. (2003). *Copyleft and dual licensing for publicly funded software development*. Retrieved July 6, 2004, from http://www.infonomics.nl/FLOSS/ papers/dual.htm

Hann, I-H., Roberts, J., Slaughter, S., & Fielding, R. (2004). *An empirical analysis of economic returns to open source participation*. Retrieved July 12, 2006, from http://www.andrew.cmu.edu/user/jroberts/Paper1.pdf

Hars, A., & Ou, W. (2001). Working for free?—Motivations of participating in open source projects. In *34ᵗʰ Annual Hawaii International Conference on System Sciences* (Vol. 7, pp. 1-7).

Haruvy, E., Prasad, A., & Sethi, S. P. (2003). Harvesting altruism in open-source software development. *Journal of Optimization Theory and Applications, 118*(2), 381-416.

Hecker, F. (1999). Setting up shop: The business of open-source software. *IEEE Software, 16*(1), 45-51.

Hertel, G., Niedner, S., & Herrmann, S. (2003). Motivation of software developers. *Research Policy, 32*(7), 1159-1177.

Kalla, R. (2006). *Eclipse as an ecosystem.* Retrieved March 3, 2006, from http://www.eclipsezone.com/eclipse/forums/t64080.rhtml

Krishnamurthy, S. (2002). *Cave or community? An empirical examination of 100 mature open source projects.* Retrieved May 4, 2007, from http://opensource.mit.edu/papers/krishnamurthy.pdf

Lakhani, K. R., & Wolf, R. G. (2003). *Why hackers do what they do: Understanding motivation effort in free/open source software projects.* Retrieved October 6, 2003, from http://opensource.mit.edu/papers/lakhaniwolf.pdf

Leiteritz, R. (2004). Open Source-Geschäftsmodelle. In R. A. Gehring & B. Lutterbeck (Eds.), *Open Source Jahrbuch 2004* (pp. 139-170). Berlin: Lehmanns Media.

Lerner, J., & Tirole, J. (2002). Some simple economics of open source. *Journal of Industrial Economics, 52*(6), 197-234.

Luthiger, B. (2006). *Spass und Software-Entwicklung. Zur Motivation von Open-Source-Programmierern.* Stuttgart: ibidem-Verlag.

Nichols, D. M., & Twidale, M. B. (2002). *Usability and open source software.* Retrieved October 4, 2003, from http://www.cs.waikato.ac.nz/~daven/docs/oss-fm.pdf

O'Mahony, S., Cela Diaz, F., & Mamas, E. (2005). *IBM and Eclipse.* Harvard: Harvard Business School.

Osterloh, M., Rota, S., & Kuster, B. (2002). *Open source software production: Climbing on the shoulders of giants.* Retrieved July 21, 2003, from http://www.iou.unizh.ch/orga/downloads/publikationen/osterlohrotakuster.pdf

Raymond, E. S. (1999). *The magic cauldron.* Retrieved July 21, 2003, from http://www.catb.org/~esr/writings/magic-cauldron/

Raymond, E. S. (2000). *Homesteading the noosphere.* Retrieved July 21, 2003, from http://www.catb.org/~esr/writings/homesteading/

Stamelos, I., Angelis, L., Oikonomou, A., & Bleris, G. L. (2002). Code quality analysis in open source software development. *Information Systems Journal, 12*(1), 43-60.

Toffler, A. (1980). *The third wave.* New York: Bantam Books.

Torvalds, L., & Diamond, D. (2001). *Just for FUN: The story of an accidental revolutionary.* New York: Harper Collins Publishers.

von Hippel, E., & von Krogh, G. (2002). *Exploring the open source software phenomenon: Issues for organization science*. Retrieved July 21, 2003, from http://opensource.mit.edu/papers/removehippelkrogh.pdf

Weber, S. (2000). *The political economy of open source software*. Retrieved May 4, 2002, from http://e-conomy.berkeley.edu/publications/wp/wp140.pdf

Weber, S. (2004). *The success of open source*. Cambridge, MA: Harvard University Press.

Weiss, D. (2005). Measuring success of open source projects using Web search engines. In M. Scotto & G. Succi (Eds.), *Proceedings of the 1st International Conference on Open Source Systems*, Genova (pp. 139-170).

Endnotes

[1]   In a study about "Fun and Software Development" (FASD), we explored the importance of fun that the open source developers enjoy on their open source engagements. The survey addressed both open source and commercial developers and was filled out by 1330 programmers from the open source area and by 114 developers working in six Swiss software companies. The surveys have been open during about two months in early summer and autumn 2004 respectively.

[2]   A few exceptions can be noted: see Gabriel and Goldman (2002), Dahlander (2004), Bonacorsi and Rossi (2005) for example.

[3]   "I'm looking for people for whom knowing they are helping humanity is as important as money" (Richard Stallman on www.gnu.org/gnu/initial-announcement.html).

Section IV

Adoption of F/OSS in Public and Corporate Environments

Chapter VIII

Understanding the Development of Free E-Commerce/E-Business Software:
A Resource-Based View[1]

Walt Scacchi, University of California, Irvine, USA

Abstract

This study examines the development of open source software supporting e-commerce (EC) or e-business (EB) capabilities. This entails a case study within a virtual organization engaged in an organizational initiative to develop, deploy, and support free/open source software systems for EC or EB services, like those supporting enterprise resource planning. The objective of this study is to identify and characterize the resource-based software product development capabilities that lie at the center of the initiative, rather than the software itself, or the effectiveness of its operation in a business enterprise. By learning what these resources are, and how they are arrayed into product development capabilities, we can provide the knowledge needed to understand what resources are required to realize the potential of free EC and EB software applications. In addition, the resource-based view draws attention to those resources and capabilities that provide potential competitive advantages and disadvantages to the organization in focus.

Introduction and Background

Many companies face a problem in determining how to best adopt and deploy emerging capabilities for e-commerce and e-business services. This study employs a *resource-based view* of the organizational system involved in developing open source EC/EB software products or application systems. This chapter examines the GNUenterprise.org (hereafter GNUe) project as a case study. The analysis and results of the case study focus attention on characterizing an array of social and technical resources the developers must mobilize and bring together in the course of sustaining their free EC/EB software development effort. Free EC/EB results from applying free software development concepts, techniques, and tools (Williams, 2002) to supplant those for open source software supporting EC and EB (cf. Carbone & Stoddard, 2001).

This study does not focus on the software functionality, operation, or development status of the GNUe free EC/EB software, since these matters are the focus of the GNUe effort, and such details can be found on that project's Web site. Similarly, it does not discuss what EC/EB application packages are being developed or their operational status, though the categories of software packages can be seen in Exhibit 1, presented later. Instead, the resource-based view (Acedo et al., 2006; Barney, 2001) that is the analytical lens employed in this chapter helps draw attention to a broader array of resources and institutionalized practices (i.e., resource-based capabilities) (Oliver, 1997) that may better characterize the socio-technical investments that provide a more complete picture of the non-monetized costs associated with the development of free/open source software (FOSS), as well as possible competitive advantages and disadvantages (Hoopes et al., 2003). Such a characterization might then eventually inform other studies that seek to identify and explain the "total costs of operations" involved in developing, deploying, and sustaining FOSS, or the commercial services that support these costs.

Case Study: The Development of Free EC/EB Software in GNUe

GNUe is an international virtual organization for software development (Crowston & Scozzi, 2002; Noll & Scacchi, 1999) based in the U.S. and Europe that is developing an enterprise resource planning (ERP) system and related EC/EB packages using only free software. One of their overarching goals is to put freedom back into "free enterprise," as seen in the overview of GNUe shown in Exhibit 1, which is taken from the project's Web site. This organization is centered about the GNUenterprise. org Web site/portal that enables remote access and collaboration. Developing the

GNUe software occurs through the portal that serves as a global information-sharing workplace and collaborative software development environment. Its paid participants are sponsored by one or more of a dozen or so companies spread across the U.S. and Europe. These companies provide salaried personnel, computing resources, and infrastructure that support this organization. However, many project participants support their participation through other means. In addition, there are also dozens of unpaid volunteers who make occasional contributions to the development, review, deployment, and ongoing support of this organization, and its software products and services. Finally, there are untold numbers of "free riders" (Olson, 1971) who simply download, browse, use, evaluate, deploy, or modify the GNUe software with little/no effort to contribute back to the GNUe community.

GNUe is a community-oriented project, as are most sustained FOSS development efforts (Scacchi, 2002a; Sharman et al., 2002; West & O'Mahony, 2005). The project started in earnest in 2000 as the result of the merger of two smaller projects both

Exhibit 1. Overview of the GNUe and its GNUe software (Source: Retrieved April 2006, from http://www.gnuenterprise.org/)

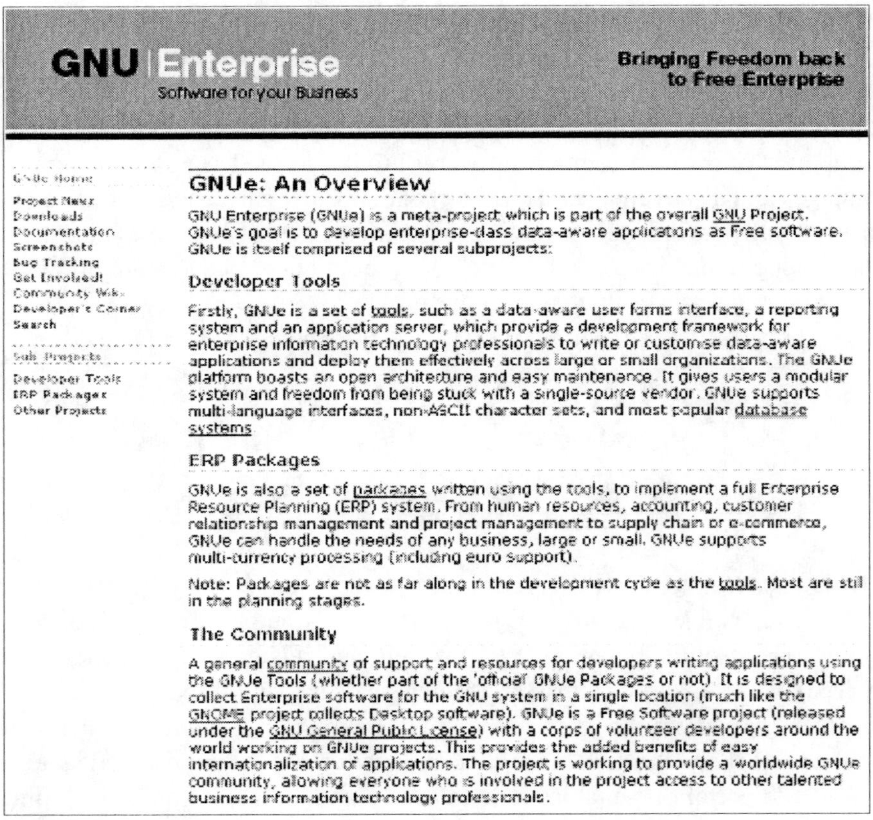

seeking to develop a free software solution for EC/EB applications. More information on the history of the GNUe project can be found on their Web site.

The target audience for the GNUe software application packages is envisioned primarily as small to mid-size enterprises (SMEs) that are underserved by the industry leaders in ERP software. These SMEs may be underserved due to the high cost or high prices that can be commanded for commercial ERP system installations. Many of these target SMEs might also be in smaller or developing countries that lack a major IT industry presence.

GNUe is a free software project affiliated with the Free Software Foundation and the European FSF. The ERP and EC/EB software modules and overall system architecture are called the GNUe software. All the GNUe software is protected using the GNU Public License (GPL) (DiBona, Ockman, & Stone, 1999; Pavlicek, 2000; Williams, 2002). This stands in contrast to the open source ERP software from Compiere[2], which depends on the use of a commercial Oracle DBMS, or other commercially-based OSS ERP project like OpenMFG.com and Openbravo.com. Thus, GNUe is best characterized as a *free software* project (Williams, 2002), rather than simply an open source software project (Feller & Fitzgerald, 2002). But many GNUe participants also accept its recognition as an open source software project, since most OSS and all free software projects employ the GPL to ensure the FOSS nature of their development activities and products.

GNUe itself is not in business as a commercial enterprise that seeks to build products and/or offer services. It is not a dot-com business, but is a "dot-org" community venture. The "business model" of GNUe is more of a pre-competitive alliance (or a "cooperative") of software developers and companies that want to both cooperate and participate in the development and evolution of free ERP and EC/EB software modules. As such, it has no direct competitors in the traditional business sense of market share, sales and distribution channels, and revenue streams.

GNUe does not represent a direct competitive threat to ERP vendors like SAP, Oracle, or JD Edwards. This will be true until these companies seek to offer low-cost, entry-level ERP or EC/EB service applications for SME customers. However, it does compete for attention, participation, independent consulting engagements, and mindshare from potential FOSS developers/users with companies like Compiere.com, OpenMFG.com, Openbravo.com, and others that seek to develop and deploy OSS for ERP applications and EC/EB service offerings that may incorporate non-free, closed source, proprietary software. In addition, since the development of the GNUe software is open for global public review and corporate assessment, it is possible that the efforts and outcomes of GNUe might influence other companies developing ERP or EC/EB software. For example, other non-free, closed source ERP software vendors may perceive competitive pressure of new system features, lower cost software products, better quality, more rapid maintenance, or modular system architectures (CW360, 2002) arising from the globally visible FOSS development efforts of GNUe.

The GNUe virtual organization is informal. There is no lead organization or prime contractor that has brought together the alliance as a network. It is more of an emergent organizational form where participants have in a sense discovered each other, and have brought together their individual competencies and contributions in a way whereby they can be integrated or made to interoperate (Crowston & Scozzi, 2002; Crowston & Howison, 2005). In GNUe, no company or corporate executive has administrative authority or resource control to determine: (a) what work will be done; (b) what the schedule will be; (c) who will be assigned to perform specified tasks; (d) whether available resources for the project are adequate, viable, or extraneous; or (e) who will be fired or reassigned for inadequate job performance. As such, there is comparatively little administrative overhead to sustain ongoing software development and community portal support activities. Instead, there is a group of core developers, secondary contributors, and casual volunteers who review and comment on what has been done (cf. Jensen & Scacchi, 2007). The participants come from different small companies or act as individuals that collectively move the GNUe software and the GNUe community forward. Thus, the participants self-organize in a manner more like a meritocracy (Fielding, 1999; Scacchi, 2004), rather than a well-orchestrated community for Web-based commerce or entertainment (Kim, 2000).

Certain kinds of software development decisions are made by "logically centralized but physically distributed" core developers (cf. Noll & Scacchi, 1999). These core developers have earned the trust, sustained their commitment of personal time and effort on the project, have been recognized as technical authorities in the project, and have achieved some degree of "geek fame" in the eyes other project participants (cf. Fielding, 1999; Pavlicek, 2000). Like other project participants and FOSS developers, the GNUenterprise core developers are expected to uphold and reiterate the freedom of expression, sharing, and learning that free, open source GNUe software represents or offers. So as core developers of GNUe software, they must reflect on how their software development decisions reflect, embody, or otherwise reproduce belief in free, open source software. On the other hand, decisions to contribute gifts of skill, time, effort, and other production resources that give rise to software, online communications, and technical peer reviews, are externalized or decentralized across a virtual organization (Bergquist & Ljundberg, 2001; Crowston & Scozzi, 2002). This decentralization of costs reduces the apparent direct cost and administrative overhead (indirect cost) of OSSD by externalization and global distribution, while sustaining something of a centralized decision-making authority. Thus, individual, corporate, and collective self-interest are motivated, sustained, and renewed in a manner accountable to the culture and community that is GNUe (cf. Monge et al., 1998).

As such, these conditions make this study unique in comparison to previous case studies of EC or EB initiatives, which generally assume the presence of a centralized administrative authority and locus of resource control common in large firms (e.g., Scacchi, 2001). But it is similar to prior FOSS case studies (e.g., Scacchi, 2002a; German,2003) that focus attention on the array of resources whose value is simultaneously both social and technical (i.e., socio-technical resources). Nonetheless, we still need a better understanding of what resource-based capabilities are brought to bear on the development and deployment of EC/EB and ERP software by GNUe. Subsequently, what follows is a description of key resources being employed throughout GNUe to develop and support the evolution of the GNUe software modules.

Analyzing the GNUe Case

This section presents an interpretive analysis of the case study, as is appropriate for the kinds of data and descriptions that have been presented and in related studies (cf. Scacchi, 2001, 2002a; Skok & Legge, 2002).

A reasonable question to ask at this point is whether GNUe is an efficient and effective enterprise, and whether its participants realize gains that outweigh their individual investments. As a FOSS development alliance and virtual enterprise, GNUe is not designed to make money or be profitable in the conventional business sense. It is, however, conceived to be able to develop and deploy complex ERP and EC/EB software modules. Companies that provide paid software developers to work on the GNUe software expect to make money from consulting, custom systems integration and deployment, and ongoing system support. These services generally accompany the installation and deployment of this kind of software. They may also just seek to acquire, use, and deploy open ERP or EC/EB applications for their own internal EB operations. Similarly, they may value the opportunity to collaborate with other firms or other highly competent ERP and EC/EB software developers (Crowston & Scozzi, 2002; Jensen & Scacchi, 2007; Monge et al., 1998). Other unpaid contributors and volunteers may also share in these same kinds of values or potential outcomes.

Can an enterprise make money from creating a complex ERP and EC/EB software suite that from the start is distributed as free, open source software? Don't ERP and EC/EB software products whose proprietary closed source alternatives from SAP and others cost upwards of a million dollars or more (Curran & Ladd, 2000; Keller & Tuefel, 1998)? Yes, closed source ERP and EC/EB systems do entail substantial acquisition, implementation, deployment, and support costs. But the purchase price of most ERP software packages and EC/EB service application may only represent 5-10% of the total cost of a sustained deployment in a customer enterprise. Subsequently, most of the financial cost of an ERP or EC/EB application deployment is in providing the installation, customization, and maintenance support services. As

FOSS in widespread use is subject to continuous improvement, the opportunity to provide ongoing support services to businesses or government agencies that rely on them will continue and grow. Thus, a FOSS project like GNUe can still serve to generate opportunities for support service providers, without the need to generate revenues from sales of their ERP and EC/EB software. Commercial vendors like IBM, RedHat, JBoss (recently acquired by RedHat), and many others offer many kinds of OSS support services to realize their revenue generation goals, so GNUe's developers have the potential to earn a living or make additional money from their FOSS development efforts.

What kinds of challenges can make the transition from EC/EB to free EC/EB problematic or motivating, and how might these problems be mitigated via OSSD? Two broad categories of challenges to free EC/EB are apparent: those involving economic conditions such as those already noted, and those denoting structural or resource-based capabilities (Acedo et al., 2006; Barney, 2001; Hoopes et al., 2003; Oliver, 1997). Here the focus is on the later, and thus start with a description of the research methods employed in this study.

Research Methods

This study of GNUe arises from a longitudinal field study spanning 2002-2006. The study employed grounded theory techniques (Glaser & Strauss, 1967; Strauss & Corbin, 1980) including axial coding and construction of comparative memoranda based on field data collected through face-to-face and email-interviews, as well as extensive collection and cross-coding of publicly available project documents and software development artifacts posted on the project's Web site. These field study methods are subsequently closely aligned with those characterized as virtual ethnography (cf. Hakken, 1999; Hine, 2000; Scacchi, 2002a) applied to software development projects (cf. Viller & Sommerville, 2000) operating over the Internet/Web as a distributed virtual enterprise (Noll & Scacchi, 1999). A diverse set of work practices and socio-technical interaction processes emerged from the codings and their comparative analysis. These include how participants in different roles express their beliefs, norms, and values, as well as how they are enacted in shaping what free software development entails (Elliott & Scacchi, 2003). These, in turn, guide technical decision-making regarding which tools to employ during development activities, as well as how globally distributed participants act through cooperation and conflict to collectively form (and re-form) GNUe as a virtual organization (Elliott & Scacchi, 2005). Finally, these practices also serve as a basis for articulating an occupational community of free software developers within the free software movement (Elliott & Scacchi, 2003, 2006). The study presented here extends and complements those just cited through a reframing of the observed practices through data coding and institutionalized patterns that characterize the socio-technical resources and resource-based capabilities that support free software development work in GNUe. Finally, the analysis employs a variety of representational notations, relational schemes, and flow diagrams (Scacchi et al., 2006) to help articulate the results that are described next.

Resources and Capabilities for Developing Free EC/EB Software in GNUe

What kinds of resources or business capabilities are needed to help make free EC/EB efforts more likely to succeed? How do these resources differ from those recommended in traditional software engineering projects? Based on what was observed in the GNUe case study, the following (unordered) set of socio-technical resources and capabilities enable the development of (a) free ERP and EC/EB software packages, as well as (b) the community that is sustaining its evolution, deployment, and refinement, though other kinds of socio-technical processes also play key roles in mobilizing these resources into capabilities supporting work practices, and these are described elsewhere (Elliott & Scacchi, 2003, 2005, 2006; Scacchi, 2005).

Personal Software Development Tools and Networking Support

In GNUe, free software developers provide their own personal computer resources (often in their homes) in order to access or participate in the project. They similarly provide their own access to the Internet, and some even host personal Web sites or information repositories. Furthermore, these free software developers bring their own choice of tools (e.g., source code compliers, diagram editors) and development methods to the GNUe community, though this seems to be common to many FOSS projects. There are few shared computing resources beyond the project's Web site, though its operation is supported in part by a company that provides a small number of programmers to work on the GNUe software. Nonetheless, the sustained commitment of personal resources helps *subsidize* the emergence and evolution of the GNUe community, its shared (public) information artifacts, and resulting free software. It also helps create recognizable shares of the free software commons (cf. Benkler, 2006; Lessig, 2005) that are linked (via hardware, software, and the Web) to the community's information infrastructure.

Beliefs Supporting FOSS Development

Why do free software developers contribute their skill, time, and effort to the development of free software and related information resources? Though there are probably many diverse answers to such a question, it seems that one such answer must account for the belief in the freedom to access, study, modify, redistribute, and share the evolving results from a FOSS development project. Without such belief, it seems unlikely that there could be "free" and "open source" software development projects (DiBona, Ockman, & Stone, 1999; Pavlicek, 2000; Williams, 2002). However, one important consideration that follows is what the consequences from such belief are, and how these consequences are put into action.

In looking across the case study data, in addition to examination of the online GNUe information resources from which they were taken (cf. Elliott & Scacchi, 2003, 2005, 2006), many kinds of actions or choices emerge from the development of free software. Primary among them in the GNUe project (and possibly other FOSS projects) are freedom of expression and freedom of choice. Neither of these freedoms is explicitly declared, assured, or protected by free software copyright (the GNU Public License, GPL) or community intellectual property rights, or end-user license agreements.[3] However, they are central tenets of free or open source modes of production and culture (Benkler, 2006; Lessig, 2005). In particular, in FOSS projects like GNUenterprise and others, these additional freedoms are expressed in choices for what to develop or work on (e.g., choice of work subject or personal interest over work assignment), how to develop it (choice of method to use instead of a corporate standard), and what tools to employ (choice over which personal tools to employ versus only using what is provided). Consider the following excerpt from an online chat provided by someone (here identified with the pseudonym, ByronC) who was an outsider to the day-to-day development activities in the GNUe project seeking to determine if free (appropriate) or non-free (inappropriate) software tools were being used to create diagrams that help document and explain the how the GNUe software is organized:

<ByronC> Hello. Several images on the Website seem to be made with non-free Adobe software. I hope I am wrong; it is quite shocking. Does anybody know more on the subject? We should avoid using non-free software at all cost, am I wrong?

Elsewhere, GNUe developers also expressed choices for when to release work products (choice of satisfaction of work quality over schedule), determining what to review and when (modulated by community ownership responsibility), and expressing what can be said to whom with or without reservation (modulated by trust and accountability mechanisms). Shared belief and practice in these freedoms of expression and choice are part of the virtual organizational culture that characterizes a community project like GNUe (Elliott & Scacchi, 2003, 2005). Subsequently, putting these beliefs and cultural resources into action continues to build and reproduce socio-technical interactions networks that enable sustained FOSS project community and the free software movement (Elliott & Scacchi, 2006; Scacchi, 2005).

Competently Skilled and Self-Organizing Software Developers

Developing complex software modules for ERP applications requires skill and expertise in the domain of EB and EC. Developing these modules in a way that enables an open architecture requires a base of prior experience in constructing open systems. The skilled use of project management tools for tracking and re-solving open issues, and also for bug reports contributing to the development of such system architecture. These are among the valuable professional skills that are mobilized, brought, or drawn to FOSS development community projects like GNUe (cf. Crowston & Scozzi, 2002; Crowston & Howison, 2005). These skills are resources that FOSS developers bring to their projects, much like any traditional software development project.

FOSS developers organize their work as a virtual organizational form that seems to differ from what is common to in-house, centrally-managed software develop-ment projects, which are commonly assumed in traditional software engineering textbooks (Sommerville, 2004). Within in-house development projects, software application developers and end-users often are juxtaposed in opposition to one another. Danziger (1979) referred to this concentration of software development skills, and the collective ability of an in-house development organization to control or mitigate the terms and conditions of system development as a "skill bureaucracy." Such software development skill bureaucracy (though still prevalent today) would seem to be mostly concerned with rule-following and rationalized decision-making, perhaps as guided by a "software development methodology" and its corresponding "interactive development environment" for software engineering.

In a decentralized virtual organization of a FOSS development community like GNUe, a "skill meritocracy" (cf. Fielding, 1999) appears as an alternative to the skill bureaucracy. In such a meritocracy, there is no proprietary software develop-ment methodology or tool suite in use. Similarly, there are few explicit rules about what development tasks should be performed, who should perform, when, why, or how. However, this is not to say there are no rules that serve to govern the project or collective action within it.

The rules of governance and control in the GNUe project are informally articu-lated but readily recognized by project participants. These rules serve to control the rights and privileges that developers share or delegate to one another in areas such as who can commit source code to the project's shared repository for release and redistribution (cf. Fogel, 1999). Similarly, rules of control are expressed and incorporated into the open source code itself in terms of how, where, and when to access system-managed data via application program interfaces, end-user interfaces, or other features or depictions of overall system architecture. But these rules may and do get changed through ongoing project development and online discourse carried out in the GNUe project's persistent online chat records.

Subsequently, GNUe project participants self-organize around the expertise, reputation, and accomplishments of core developers, secondary contributors, and tertiary reviewers and other volunteers. This, in turn, serves to help them create a logical basis for their collective action in developing the GNUe free software (cf. Olson, 1971). Thus, there is no assumption of a communal or egalitarian authority or utopian spirit. Instead what can be seen is a pragmatic, continuously-negotiated order that tries to minimize the time and effort expended in mitigating decision-making conflicts while encouraging cooperation through reiterated and shared beliefs, values, and norms (Elliott & Scacchi, 2005; Espinosa et al., 2002).

In GNUe, participants nearer the core have greater control and discretionary decision-making authority, compared to those further from the core (cf. Lave & Wenger, 1991; Crowston & Howison, 2006). However, realizing such authority comes at the price of higher commitment of personal resources described above. For example, being able to make a decision stick or to convince other community participants as to the viability of a decision, advocacy position, issue, or bug report also requires time, effort, communication, and creation of project content to substantiate such an action. Such articulation can be seen in the daily records of the project's online chat archive. The authority brought about through such articulation also reflects developer experience as an interested end-user of the software modules being developed. Thus, developers possessing and exercising such skill may be intrinsically motivated to sustain the evolutionary development of their free open source ERP and EC/EB software modules, so long as they are active participants in the GNUe project community.

Discretionary Time and Effort of Developers

Are FOSS developers working for "free," or for advancing their career and professional development? Most of the core GNUe software developers have "day jobs" as software developers or consultants in companies, but few of these jobs specifically focus on the development of FOSS. So developing free software in the GNUe project is supported only in part for some of its core developers. Elsewhere, the survey results of Hars and Ou (2002) and others (Lerner & Tirole, 2000; Hann et al., 2002) suggest there are many personal and professional career-oriented practices for why participants will contribute their own personal (unpaid) time and effort to the sometimes difficult and demanding tasks of software development. What we have found in GNUe appears consistent with the cited observations. These practices that help motivate action include self-determination, peer recognition, community identification, and self-promotion, as well as belief in the inherent value of free software (cf. DiBona, Ockman, & Stone, 1999; Pavlicek, 2000; Williams, 2002).

In the practice of self-determination, no one has the administrative authority to tell a project member what to do, when, how, or why. GNUe developers can choose to work on what interests them personally, though their choices are limited to features or functions relevant to the ERP or EC/EB packages (or support libraries) they are developing. GNUe developers, in general, work on what they want, when they want, though the core developers do routinely connect to the project's chat room as a way to show up for work and to be visible to others.. However, they remain somewhat accountable to the inquiries, reviews, and messages of others in the community, particularly with regard to software modules or functions for which they have declared their responsibility to maintain or manage as a core developer.

In the practice of peer recognition, a GNUe developer becomes recognized as an increasingly valued community contributor as a growing number of their contributions make their way into the core software modules (Benkler, 2006; Bergquist & Ljundberg, 2001). In addition, nearly two-thirds of FOSS developers work on 1-10 additional software projects (Hars & Ou, 2002; Madey et al., 2005), which also reflect a growing social network of alliances across multiple software development projects (cf. Monge et al., 1998; Scacchi, 2005). The project contributors who span multiple free or non-free software project communities (identified as "linchpin developers" by Madey et al., 2005) serve as "social gateways" that increase the GNUe community's mass (Marwell & Oliver, 1993), as well as affording opportunities for inter-project software composition, bricolage, and interoperation (Jensen & Scacchi, 2005). For example, some of the core developers chose to import and integrate a free project reporting system (previously in use in other software projects) to help keep track of outstanding GNUe software bugs, as well as who is working on what.

In building community identification, GNUe project participants build shared domain expertise and identify who is expert in knowing how to do what (cf. Ackerman & Halverson, 2000). Interlinked information on the project's Web site, project development artifacts, and persistent online chat messages help point to whom the experts and core contributors are within the project's socio-technical interaction network (Scacchi, 2005).

In self-promotion, GNUe project participants communicate and share their experiences, perhaps from other application domains or work situations, about how to accomplish some task, or how to develop and advance through one's career. Being able to move from the project periphery towards the center or core of the development effort requires not only the time and effort of a contributor, but also the ability to communicate, learn from, and convince others as to the value or significance of the contributions (cf. Jensen & Scacchi, 2007; Lave & Wegner, 1991). This is necessary when a participant's contribution is being questioned in open project communications, not incorporated (or "committed") within a new build version, or rejected by a vote of those already recognized as core developers (cf. Fielding, 1999).

The last source of discretionary time and effort observed in GNUe is found in the freedoms and beliefs in FOSSD that are shared, reiterated, and put into observable interactions. If a community participant fails to sustain or reiterate the freedoms and beliefs codified in the GPL, then it is likely the person's technical choice in the project may be called into question (Elliott & Scacchi, 2003, 2005), or the person will leave the project and community. But understanding how these freedoms and beliefs are put into action points to another class of resources (sentimental resources) that must be mobilized and brought to bear in order to both develop FOSS systems and the global communities that surround and empower them. Social values that reinforce and sustain the project community, and technical norms regarding which software development tools and techniques to use (e.g., avoid the use of "non-free" software), are among the sentimental resources that are employed when participants seek to influence the choices that others in the project seek to uphold.

Trust and Social Accountability Mechanisms

Developing complex software modules for ERP and EC/EB applications requires trust and accountability among GNUe project participants. Though trust and accountability in a FOSS project may be invisible resources, ongoing software and community development work occur only when these intangible resources and mechanisms for social control are present (cf. Gallivan, 2001; Hertzum, 2002).

The intangible resources of trust and accountability in GNUe arise in many forms. They include assuming ownership or responsibility of a community software module, voting on the approval of an individual action or contribution to community software (Fielding, 1999), shared peer reviewing of developer work products (DiBona, Ockman, & Stone, 1999; Benkler, 2006), and by contributing gifts (Bergquist & Ljundberg, 2001) that are reusable and modifiable public goods (Olsen, 1971; Samuelson, 1954; Lessig, 2005). They also exist through the community's recognition of a core developer's status, reputation, and geek fame (Pavlicek, 2000). Without these attributions, GNUe developers may lack the credibility they need to bring conflicts over how best to proceed to some accommodating resolution. Finally, as the GNUe project has been sustained (though with turnover) for over five years in terms of the number of contributing developers, end-users, and external sponsors, then GNUe's socio-technical mass (i.e., web of interacting resources) has become sufficient to ensure that individual developer trust and accountability to the project community are sustained and evolving (Marwell & Oliver, 1993).

Thus, the GNUe participants rely on mechanisms and conditions they have created for gentle but sufficient social control that help constrain the overall complexity of the project. These constraints act in lieu of an explicit administrative authority or project management regime that would schedule, budget, staff, and control the project's development trajectory with varying degrees of administrative authority and technical competence, as would be found in a traditional software engineering project (cf. Sommerville, 2004).

Free Open Source Software Development Informalisms

Software informalisms (Scacchi, 2002a) are the information resources and artifacts that participants use to describe, proscribe, or prescribe what's happening in a FOSSD project. They are informal narrative resources (or online document genres, cf. Kwansik & Crowston, 2005) that are comparatively easy to use, and immediately familiar to those who want to join the community project. However, the contents they embody require extensive review and comprehension by a developer before core contributions can be made. The most common informalisms used in GNUe include: (1) community communications and messages within project e-mail, (2) threaded message discussion forums or group blogs, (3) project news postings, (4) community digests, and (5) instant messaging or Internet relay chat. They also include (6) scenarios of usage as linked Web pages, (7) how-to guides, (8) to-do lists, (9) FAQs and other itemized lists, and (10) project Wikis, as well as (11) traditional system documentation and (12) external publications. Free software (13) community property licenses also help to define what software or related project content are protected resources, so that they can subsequently be shared, examined, modified, and redistributed. Finally, (14) open software architecture diagrams, (15) intra-application functionality realized via scripting languages like Perl and PhP, and the ability to either employ (16) plug-in components or (17) integrate software modules from other OSSD efforts, are all resources that are used informally, where or when needed according to the interests or actions of project participants.

All of the software informalisms are found or accessed from (18) project related Web sites or portals (see Exhibit 1). These Web environments are where most FOSS software informalisms can be found, accessed, studied, modified, and redistributed (Scacchi, 2002a). A Web presence helps make visible the GNUe community's information infrastructure and the array of information resources that populate it. These include FOSS development community project Web sites (e.g., SourgeForge.net, Savanah.org), community software Web sites (PhP-Nuke.org), as well as (19) embedded project source code Webs (directories), (20) project repositories (CVS) (Fogel, 1999), (21) software bug reports and (22) issue-tracking databases (called DCL in GNUe).

Together, these two dozen or so types of software informalisms constitute a substantial yet continually evolving web of informal, semi-structured, or processable information resources within GNUe. This web results from the hyperlinking and cross-referencing that interrelate the contents of different informalisms together. Subsequently, these FOSS informalisms are produced, used, consumed, or reused within GNUe. They also serve to act as both a distributed virtual repository of FOSS project assets, as well as the continually adapted distributed knowledge base through which project participants in GNUe evolve what they know about the software systems they develop and use.

FOSSD Capability Enabling Free, Open ERP, and EC/EB Systems

The array of social, technological, and informational resources that enable a FOSS development project like GNUe is substantial. However, they differ in kind and form from the traditional enterprise resources that are provided to support proprietary, closed source software systems. These traditional software engineering resources are money (budget), time (schedule), skilled (salaried) development staff, project managers (administrative authority), quality assurance (QA) and testing groups, documentation writers, computer hardware and network maintainers, and others (cf. Sommerville, 2004). Free software projects like GNUe seem to get by with comparatively small amounts of money, though subsidies of various kinds and sources are present and necessary. They also get by without explicit schedules, though larger projects may announce target release dates, as well as (partially) order which system functions or features will be included in some upcoming versions, for some target releases. Further, they get by without a rule-making and decision-making authority of corporate project managers or enterprise executives, who may or may not be adept at empowering, coaching, or rewarding development staff to achieve corporate software development goals. Instead, in GNUe, participants rely on an implicit but frequently recited regime of beliefs, values, and norms that help organize cooperative activity and rationalize conflict mitigation (Elliott & Scacchi, 2003, 2005). The remaining resources are provided within a free software development effort via subsidies, sponsorship, or volunteer effort.

Thus, the resources for free software development efforts are different in kind, and in how they are arrayed and brought to bear when compared to a traditional software engineering effort. Free software project resources are not mobilized, allocated, or otherwise brought to bear in the manner traditional to the development of proprietary, closed source software systems. Hopefully, it should be clear that the differences being highlighted are not based simply on a comparison of functionality or features visible in the development or use of open vs. close source software products. As such, the resource-based capability for developing free software packages for ERP and EC/EB applications is different, though not necessarily more or less costly.

Discussion and Conclusion

Many questions about free software development remain unanswered or unexamined by this study. For example, it is unclear whether there must be a critical mass of salaried software developers whose job includes development or support of GNUe software, and if so, how many software developers this entails. Over the four years in the study of GNUe, the number and composition of core software developers has changed, partly in response to their changing interests and work situations. The project has not grown to the point where commercialization of the GNUe software has become an imperative or new venture start-up opportunity, as has happened with OSS ERP projects of Compiere, OpenMFG, and Openbravo. Thus, it is unclear whether GNUe will become a viable enterprise capable of hiring a professional or full-time staff, as well as engaging in contracted provision of installation, customization, and support services that normally accompany ERP and EC/EB software packages. Further comparative study of other free software projects and their approach for commercialization are needed. However, it does seem clear that GNUe has managed to sustain itself as a viable ongoing enterprise that continues to develop and sustain free ERP and EC/EB software packages, which its developers use and deploy in their day jobs or consulting practices.

Beyond this, three conclusions can be drawn from the study, data, and analysis presented in this report. First, this chapter identifies many types of socio-technical resources and resource-based capabilities for free EC/EB that may explain/predict (a) what's involved, (b) how it works, or (c) what conditions may shape the longer-term success or failure of such efforts. In simple terms, these resources include time, skill, effort, belief, personal and corporate subsidies, and community building on the part of those contributing as developers and users of free EC/EB systems and techniques. Of these, *belief* in the freedoms that open source system development embraces, including freedom of choice and freedom of expression (Elliott & Scacchi, 2003, 2005, 2006) appears central. Such belief in turn enables and affords the ongoing commitment, development, and articulation of a web of social, technological, and information resources that sustain a free software project, without the traditional administrative and financial resources found in traditional software development enterprises. Developers and users who believe in the promise and potential of free ERP or EC/EB packages are willing to allocate (or volunteer) their time and apply their skills to make the effort of developing or using open source systems a viable and successful course of action. Thus, companies seeking to invest in or exploit free EC/EB techniques or systems must account for how it can most effectively cultivate a free software culture, belief system, and community of practice, as part of their strategic choice.

Second, this is the first study to employ a resource-based view of a FOSS development project. The resource-based view of organizational capability and competitive advantage is the dominant analytical lens employed in studies of organizational strategy and strategic management (cf. Acedo et al., 2006; Barney, 2001). Why should people interested in FOSS development practices be concerned or interested in such a strategic perspective? Many reasons might be cited in support, but attention here can be drawn to determining whether free software systems and development methods offer sustained or differentiated advantages over traditional software engineering approaches applied to the development of close source, proprietary (non-free) software systems. If there are advantages that can be traced to the resource arrangements found in free software projects like GNUe, then these would be noteworthy findings, as well as a possible basis for further exploration and theorizing. Accordingly, in the GNUe case, resources like personal computing tools that help subsidize the development effort, beliefs that provide a cultural basis for making decisions about technical choices, trust and social accountability, discretionary software development work times, and the preferred use of software informalisms instead of software engineering formalisms like "requirements specifications" and "project management plans" all differentiate the practice of free software development from that advocated in traditional software engineering textbooks (e.g., Sommerville, 2004).

Last, this study links free software with ERP and EC/EB. No prior case studies of ERP or EC/EB systems have identified or addressed whether or how free software (or open source software) methods might be used to develop or integrate EC/EB software packages, at least beyond the use of OSS Web servers or Web site content management systems (Carbone & Stoddard, 2001). Thus, there is an opportunity for firms to begin considering whether these results merit timely consideration or exploratory investments in free software or OSS. For example, companies offering consumer products or high value, information technology-based products and services may begin to consider whether free EC/EB capabilities that offer lower purchase prices, lower total cost of ownership, and higher quality (Scacchi, 2002b) represent new market entry or new product differentiation opportunities. Similarly, companies may find that free/open source software represents a new, highly innovative approach to software product or application system development that marries the best capabilities from both private investment and collective action (von Hippel & von Grogh, 2003; Olson, 1971).

References

Acedo, F. J., Barroso, C., & Galan, J. L. (2006). The resource-based theory: Dissemination and main trends. *Strategic Management J.*, *27*(7), 621-636.

Ackerman, M., & Halverson, C. (2000, January). Reexamining organizational memory, *Communications ACM*, *43*(1), 59-64.

Barney, J. B. (2001). Resource-based theories of competitive advantage: A ten-year retrospective on the resource-based view. *J. Management*, *27*(6), 643-650.

Benkler, Y. (2006). *The wealth of networks*. New Haven, CT: Yale University Press.

Bergquist, M., & Ljungberg, J. (2001). The power of gifts: Organizing social relationships in open source communities. *Information Systems J.*, *11*(4), 305-320.

Carbone, G., & Stoddard, D. (2001). *Open source enterprise solutions: Developing an e-business strategy.* New York: John Wiley and Sons, Inc.

Crowston, K., & Howison, J. (2006). Hierarchy and centralization in free and open source software team communications. *Knowledge, Technology and Policy, 18*(4), 65-85, Winter.

Crowston, K., & Scozzi, B. (2002). Open source software projects as virtual organizations: Competency rallying for software development. *IEE Proceedings—Software, 149*(2), 3-17.

Curran, T. A., & Ladd, A. (2000). *SAP R/3 business blueprint: Understanding enterprise supply chain management* (2nd ed.). Upper Saddle River, NJ: Prentice-Hall.

CW360. (2002, June 12). JD Edwards pushes modular ERP. *ComputerWeekly*.

Danziger, J. (1979). The skill bureaucracy and intraorganizational control: The case of the data-processing unit. *Sociology of Work and Occupations, 21*(3), 206-218.

DiBona, C., Ockman, S., & Stone, M. (1999). *Open sources: Voices from the open source revolution.* Sebastopol, California: O'Reilly Press.

Elliott, M., & Scacchi, W. (2003, November 21-30). Free software developers as an occupational community: Resolving conflicts and fostering collaboration. In *Proc. ACM Intern. Conf. Supporting Group Work (Group'03)*, Sanibel Island, FL.

Elliott, M., & Scacchi, W. (2005). Free software development: Cooperation and conflict in a virtual organizational culture, in S. Koch (Ed.). *Free/open source software development* (pp. 152-172). Pittsburgh, PA: Idea Publishing.

Elliott, M., & Scacchi, W. (2006). Mobilization of software developers: The free software movement (submitted for publication).

Espinosa, J. A., Kraut, R. E., Slaughter, S. A., Lerch, J. F., Herbsleb, J. D., et al. (2002, December). Shared mental models, familiarity, and coordination: A multi-method study of distributed software teams. *Intern. Conf. Information Systems*, Barcelona, Spain (pp. 425-433).

Feller, J., & Fitzgerald, B. (2002). *Understanding open source software development.* NY: Addison-Wesley.

Fielding, R. (1999). Shared leadership in the Apache project, *Communications ACM, 42*(4), 42-43, 1999.

Fogel, K., (1999). *Supporting open source development with CVS.* Scottsdale, AZ: Coriolis Press.

Gallivan, M. (2001). Striking a balance between trust and control in a virtual organization: A content analysis of open source software case studies. *Information Systems J., 11*(4), 277-304.

German, D. (2003). The GNOME project: A case study of open source, global software development. *Software Process—Improvement and Practice, 8*(4), 201-215.

Glaser, B., & Strauss, A. (1967). *The discovery of grounded theory: Strategies for qualitative research.* Chicago: Aldine Publishing.

Hann, I-H., Roberts, J., Slaughter, S. L., & Fielding, R. (2002, May). Why do developers contribute to open source projects? First evidence of economic incentives. In *Proc. 2nd Workshop on Open Source Software Engineering*, Orlando, FL.

Hars, A., & Ou, S. (2002). Working for free? Motivations for participating in open-source projects. *Intern. J. Electronic Commerce, 6*(3), 25-39.

Hakken, D. (1999). *Cyborgs@Cyberspace? An ethnographer looks at the future.* London: Routledge.

Hertzum, M. (2001). The importance of trust in software engineers' assessment and choice of information sources. *Information and Organization, 12*(1), 1-18.

Hine, C.M. (2000). *Virtual ethnography.* Newbury Park, California: Sage Publications. .

Hoopes, D. G., Madsen, T. L., & Walker, G. (2003). Why is there a resource-based view? Toward a theory of competitive heterogeneity. *Strategic Management J., 24*(10), 889-902.

Jensen, C., & Scacchi, W. (2005). Process modeling across the Web information infrastructure. *Software Process—Improvement and Practice, 10*(4), 255-272.

Jensen, C., & Scacchi, W. (2007). Role migration and advancement processes in OSSD projects. In *Proceedings of the 29th Inter. Conference on Software Engineering* (to appear). Minneapolis, MN: ACM Press.

Keller, G., & Teufel, T. (1998). *SAP R/3 process oriented implementation: Iterative process prototyping* (A. Weinland, Trans.). Harlow, England: Addison Wesley Longman.

Kim, A.J. (2000). *Community building on the Web: Secret strategies of successful online communities.* Peachpit Press.

Kwansik, B., & Crowston, K. (2005). Introduction to the special issue: Genres of digital documents. *Information, Technology and People, 18*(2).

Lave, J., & Wenger, E. (1991). *Situated learning: Legitimate peripheral participation.* Cambridge, UK: Cambridge University Press.

Lerner, J., & Tirole, J. (2002). Some simple economics of open source. *Journal of Industrial Economics*, 52.

Lessig, L. (2005). *Free culture: The nature and future of creativity.* New York: Penguin Press.

Madey, G., Freeh, V., & Tynan, R. (2005). Modeling the F/OSS community: A quantative investigation. In S. Koch (Ed.). *Free/open source software development* (pp. 203-221). Hersey, PA: Idea Group Publishing.

Marwell, G., & Oliver, P. (1993). *The critical mass in collective action: A micro-social theory.* Cambridge University Press.

Monge, P. R., Fulk, J., Kalman, M. E., Flanagin, A. J., Parnassa, C., & Rumsey, S. (1998). Production of collective action in alliance-based interorganizational communication and information systems. *Organization Science, 9*(3), 411-433.

Noll, J., & Scacchi, W. (1999, February). Supporting software development in virtual enterprises. *Journal of Digital Information, 1*(4).

Oliver, C. (1997). Sustainable competitive advantage: Combining institutional and resource-based views. *Strategic Management J., 18*(9), 697-713.

Olson, M. (1971). *The logic of collective action.* Cambridge, MA: Harvard University Press.

Pavlicek, R., (2000). *Embracing insanity: Open source software development.* Indianapolis, Indiana: SAMS Publishing.

Samuelson, P. (1954). The pure theory of public expenditure. *Review of Economics and Statistics, 36,* 387-390.

Scacchi, W. (2001). Redesigning contracted service procurement for Internet-based electronic commerce: A case study. *J. Information Technology and Management, 2*(3), 313-334.

Scacchi, W. (2002a). Understanding the requirements for developing open source software systems, *IEE Proceedings—Software, 149*(2), 24-39.

Scacchi, W. (2002b). Is open source software development faster, better, and cheaper than software engineering? In *Proceedings 2nd Workshop on Open Source Software Engineering,* Orlando, FL.

Scacchi, W. (2004, January/February). Free/open source software development practices in the game community. *IEEE Software, 21*(1), 59-67.

Scacchi, W. (2005). Socio-technical interaction networks in free/open source software development processes. In S. T. Acuña & N. Juristo (Eds.)., *Software process modeling* (pp. 1-27). New York: Springer Science+Business Media Inc.

Scacchi, W., Jensen, C., Noll, J., & Elliott, M. (2006). Multi-modal modeling: Analysis and validation of open source software requirements processes. *Int. Journal of Information Technology and Web Eng., 1*(3), 49-63.

Sharman, S., Sugurmaran, V., & Rajagopalan, B. (2002). A framework for creating hybrid-open source software communities. *Information Systems J., 12*(1), 7-25.

Skok, W., & Legge, M. (2002). Evaluating enterprise resource planning (ERP) systems using an interpretive approach. *Knowledge and Process Management, 9*(2), 72-82.

Sommerville, I. (2004). *Software engineering* (7th ed.). New York: Addison-Wesley.

Strauss, A., & Corbin, J. (1980). *Basics of qualitative research: Techniques and procedures for developing grounded theory.* Newbury Park, CA: Sage.

Viller, S., & Sommerville, I. (2000). Ethnographically informed analysis for software engineers. *Intern. J. Human-Computer Studies, 53,* 169-196.

von Hippel, E., & von Krogh, G. (2003). Open source software and the "private-collective" innovation model: Issues for organization science. *Organization Science, 14*(2), 209-223.

West, J., & O'Mahony, S. (2005). Contrasting community building in sponsored and community founded open source projects. In *Proc. 38th Hawaii Intern. Conf. Systems Sciences,* Waikola Village, HI.

Williams, S. (2002). *Free as in freedom: Richard Stallman's crusade for free software.* Sebastopol, California: O'Reilly Books.

Endnotes

[1]   The research described in this report was supported by grants from the U.S. National Science Foundation Industry/University Research Cooperative CRITO Consortium; the National Science Foundation #0083075, #0205679, #0205724, #0350754, and # 0534771; and the Defense Acquisition University by contract N487650-27803. No endorsement implied.

[2]   Compiere.com is a software product development community that is building an open source software ERP system that requires the use of Oracle. It is not, however, free software, as in "freedom" software (Williams, 2002). Compiere.com, however, claims more than 500K copies of its software have been downloaded or installed, making it the most widely deployed ERP system in the world, whether as a proprietary or FOSS-based offering.

[3]   End-user license agreements (EULAs), associated with probably all software, often seek to declare "freedom from liability" from people who want to use licensed software for intended or unintended purposes. But liability freedom is not the focus here.

Chapter I

The Migration of Public Administrations Towards Open Source Desktop Software:
Recommendations from Research and Validation through a Case Study

Kris Ven, University of Antwerp, Belgium

Dieter Van Nuffel, University of Antwerp, Belgium

Jan Verelst, University of Antwerp, Belgium

Abstract

Several public administrations (PA) have expressed an increasing interest in open source software in the past few years and are currently migrating to open source software on the desktop. Given the large impact such a migration has on the organization, there is a need for learning from the experiences of previous migrations.

In this chapter, we deduct a number of recommendations and lessons learned from previous research conducted on the migration of PAs to open source desktop software. Next, we describe a case study on the migration of the Brussels-Capital Region towards OpenOffice.org, and compare their experiences to these recommendations. In general, our results are quite consistent with previous findings, but also indicate that additional research is still required in order to create a set of best practices—based on empirical research—for the migration towards open source software on the desktop.

Introduction

In the past few years, open source software has become a viable solution for organizations, and is being increasingly adopted. This increased popularity has been enabled by the fact that open source vendors (e.g., RedHat and SUSE) and traditional software vendors (e.g., IBM and HP) provide reliable support for open source solutions. Studies indicate, however, that organizations are primarily using open source software for server applications (see e.g., Dedrick & West, 2003; Lundell, Lings, & Lindqvist, 2006; Ven & Verelst, 2006; Wichmann, 2002). This can be explained by at least two factors. First, open source software has a strong tradition in developing server-side applications. Given this background, most open source projects are situated in horizontal domains such as Internet applications, developer tools, and technical tools (Fitzgerald, 2005). Thanks to the maturity level of most well-known open source server software (e.g., Apache and Linux), these packages are widely diffused through organizations. Successful open source software for the desktop has surfaced only recently with applications such as OpenOffice.org, Mozilla Firefox, and Mozilla Thunderbird. Second, a migration towards open source software on servers is far less disruptive for members of an organization than a migration at the desktop. In case a Web server running Microsoft IIS is replaced by the Apache Web server, or the operating system for an ERP system is changed from Unix to Linux, end users in the organization will not (or hardly) be affected by this change. A migration from Microsoft Office to OpenOffice.org will, however, affect all end users in an organization, possibly even impacting productivity.

Recently, there has been an increased interest in migrations towards open source software on the desktop. Interestingly, this trend is primarily driven by public administrations (PA). In fact, PAs can be considered pioneers in the adoption of open source desktop software. At first sight, this is actually quite remarkable. In the past, it was frequently assumed that PAs are restricted by their organizational structure, and thus limited in their innovative behavior (Nye, 1999; Moon & Bretschneider,

2002). Other studies have found PAs to be surprisingly innovative with respect to certain innovations (Bretschneider & Wittmer, 1993; Moon & Bretschneider, 2002). The use of information technology is currently considered to be an opportunity for PAs to improve their efficiency, as illustrated by the large number of recent e-government initiatives. There are two important drivers for the adoption of open source desktop software in PAs. First, it has been suggested that PAs should be conscious of their IT expenses, to make efficient use of taxpayers' money (see e.g., Applewhite, 2003; Brink, Roos, Weller, & van Belle, 2006; Fitzgerald & Kenny, 2003; Waring & Maddocks, 2005). Since the license costs for open source software are either absent or at least lower than for proprietary software, open source software has often been touted as a means for reducing overall software expenses. Since the number of desktop licenses is far greater than the number of server licenses, cost savings on the desktop may be considerably larger. Second, some authors argue that PAs should use open standards in their communication with citizens, to avoid that citizens need to buy a commercial product for communicating with the PA (see e.g., Applewhite, 2003; Kovács, Drozdik, Zuliani, & Succi, 2004b; Rossi, Scotto, Sillitti, & Succi, 2005). Other reasons for the adoption of open source software by PAs include supporting the local economy, increased flexibility and avoiding vendor lock-in (see e.g., Drozdik, Kovács, & Kochis, 2005; Kovács et al., 2004b; tOSSad, 2006; Waring & Maddocks, 2005).

Notwithstanding the advantages that open source software can offer, the migration towards open source software on the desktop will be disruptive for most users within a PA. Hence, special attention should be paid to planning the migration in order to minimize discomfort and disruptions for end users. Although the migration towards open source desktop software is a relatively new phenomenon, a number of academic studies have already described case studies of migrations of PAs towards open source software (mainly OpenOffice.org and Linux). As a result, some lessons can be learned from these migrations. Due to the exploratory nature of this research, results of these migrations are rather fragmented.

The purpose of this chapter is to integrate the recommendations from various studies on the migration of PAs towards open source software on the desktop. Subsequently, we will report on the migration of the Government of the Brussels-Capital Region towards OpenOffice.org and compare their approach to the recommendations made in previous reports. The chapter is structured as follows. First, we will provide a brief background on migrations undertaken by PAs in Europe, and the academic literature on this topic. Next, we will derive a number of recommendations with respect to the migration towards open source software on the desktop. In the fourth section, we will present the migration of the Brussels-Capital Region and compare our results to the recommendations made in previous research. Finally, we will discuss a number of implications for practice and research, and will offer our conclusions.

Related Initiatives and Studies

Open standards and open source software are increasingly used by PAs in Europe. In fact, it has been suggested that PAs will be one of the driving forces behind open source software in Europe in the next few years (González-Barahona & Robles, 2005). Indeed, several initiatives for studying the adoption of open source software have been taken at different levels in the European administration (i.e., the European Community, national governments, regional governments, and municipalities). In fact, most European countries have initiated programs to study the advantages and drawbacks of open source software in the PA (Canonico, 2005; González-Barahona & Robles, 2005), or have created policies with respect to the use of open source software[1]. The interest of the European Commission (EC) in open source software dates back to 1999, when the European Working Group on Libre Software[2] was founded. Since then, there has been increased commitment from the EC to open source software. Most important in this regard is the IDA (Interchange of Data between Administrations) program, and its successor IDABC[3] (Interoperable Delivery of European eGovernment Services to public Administrations, Businesses and Citizens). The aim of the program is to develop cross-European e-government services towards citizens and businesses. As part of the IDA program, the eEurope 2005 Action Plan[4] was written. The document states that to ensure interoperability, open standards will be used in e-government services. Furthermore, the use of open source software will be encouraged. Today, the IDABC Web site offers much information on migrations towards open source software in PAs[5]. Similarly, several research projects on the adoption and use of open source software are funded by the EC (e.g., COSPA[6], FLOSSPOLS[7], Calibre[8] and tOSSad[9]).

In Europe, several PAs have already migrated—or are planning to migrate—to open source software, including on the desktop. Several migrations of PAs are known in Austria, the Czech Republic, Finland, France, Germany, Spain, the Netherlands, and the United Kingdom[10]. Although these national initiatives vary from country to country, many of the PAs seem to at least investigate the use of open source software (González-Barahona & Robles, 2005). An illustrative sample of these initiatives is listed in Table 1. Many other initiatives are however known, especially concerning the use of open source software on servers. An example of a famous successful migration is found in Extremadura in Spain. In order to increase the IT literacy in this region—but faced with a limited budget—it was decided to base the project on open source software. This resulted in the creation of a custom Linux distribution, called *gnuLinEx*. Originally, this distribution was intended for classroom use, but is used nowadays in PAs as well (Vaca, 2005). On the other hand, some of these migrations fail—or are delayed by various problems—as illustrated with the migration of the city of Munich[11] in Germany.

Table 1. Migrations of public administrations towards open source desktop software

Country	Projects
Spain	*Extramadura region:* Project to increase connectivity and IT literacy of the region. A custom Linux distribution, gnuLinEx (http://www.linex.org), was created.
France	*Gendarmerie Nationale:* Migration towards OpenOffice.org and Mozilla Firefox on ± 80,000 desktops.
	City of Paris: Migration towards OpenOffice.org and Mozilla Firefox on 17,000 desktops.
The Netherlands	*City of Haarlem:* Migration of 2,000 desktops to OpenOffice.org.
	City of Amsterdam: Currently conducting pilots to investigate the feasibility of migrating towards OpenOffice.org.
	City of Groningen: Decided to migrate 3,650 desktops to OpenOffice.org.
Germany	*City of Munich:* Migration of 14,000 desktops to Linux.
Austria	*City of Vienna:* Migration of 7,500 desktops to OpenOffice.org and Linux. A custom Linux-distribution, Wienux (http://www.wien.gv.at/ma14/wienux.html), was created.

Several academic studies have been devoted to investigating the difficulties encountered in migrating towards open source software on the desktop and to obtaining an overview of the use of open source software by PAs. An overview of these studies is shown in Table 2. Most of these studies have used a qualitative, case study-based approach to study the migration of various public administrations. Their aim was to investigate the feasibility of the transition, describe the migration itself, and to highlight any difficulties experienced during the transition, as well as recommend solutions. Some authors have combined the qualitative approach with an experiment to investigate the usage patterns of OpenOffice.org in comparison with Microsoft (MS) Office, to provide quantitative data on the migration (Rossi et al., 2005; Rossi, Russo, & Succi, 2006).

An important study in this field is the recent FLOSSPOLS study (Ghosh & Glott, 2005), that conducted a large-scale survey among 955 PAs in 13 European countries. The study provides insight into the perceived advantages of and barriers to the use of open source software. The study also provides an overview of the extent and type of open source software that is being used by European PAs. Results show that 49% of PAs intentionally use open source software. Interestingly, many PAs seem to be unaware of their use of open source software. In about 29% of the cases, open source software is being used without the respondent being aware of the fact that the software is open source. Results further show that half of the respondents

Table 2. Studies on the adoption of open source software by public administrations

Study	Research design
Wichmann (2002)	Large-scale survey as part of the FLOSS project to study the use of open source software in organizations and PAs in Germany, UK, and Sweden.
Russo et al. (2003)	Presents the results of a trial migration performed by the Province of Bolzano-Bozen (Italy) in 10 local PAs.
Fitzgerald and Kenny (2003)	Describes the migration of an Irish (public sector) hospital towards open source software.
Zuliani and Succi (2004b)	Reports on some lessons learned on the migration towards OpenOffice.org of 60 PAs in the Province of Bolzano-Bozen (Italy).
Kovács et al. (2004b); Kovács, Drozdik, Zuliani, and Succi (2004a)	Presents an overview of the advantages and challenges in migrating towards open standards and open source software in the PA.
Zuliani and Succi (2004a)	Provides the results of a migration towards OpenOffice.org of 60 PAs in Bolzano-Bozen (Italy).
Ghosh and Glott (2005)	Follow-up survey of the FLOSS project, investigating the use of open source software in PAs in 13 European countries.
Drozdik et al. (2005)	Investigates the risks involved in migrating desktops completely to open source software (i.e., OpenOffice.org and Linux), based on a PA in Törökbálint (Hungary).
Rossi et al. (2005)	Reports on an experiment on the transition from MS Office to OpenOffice.org, studying the use of OpenOffice.org throughout 32 weeks.
Waring and Maddocks (2005)	Reports on advantages and disadvantages of open source software for the UK public sector, including the results of eight case studies.
COSPA (2005)	Reports on the experiences of the migrations in seven European PAs, conducted as part of the COSPA project.
Rossi et al. (2006)	Reports on an experiment in a PA to compare the use of MS Office and OpenOffice.org documents after migrating towards OpenOffice.org.
Brink et al. (2006)	Reports on three case studies in South African organizations who have migrated towards open source software on the desktop. Two of the cases are situated in the public sector.
Jashari and Stojanovski (2006)	Survey of 14 municipalities in Macedonia to highlight the challenges and obstacles to use open source software.

would find an increase in open source software usage useful. The use of open source software is still mainly focused on servers, with only about 20% of the organizations using OpenOffice.org to some degree (Ghosh & Glott, 2005). The use of open source software on the desktop seems, however, to be somewhat higher than three

Table 3. Recommendations and lessons learned from previous research

---

1. Analysis and Preparation

(a) Planning

- Prepare well for the migration by performing proper analysis and planning (Brink et al., 2006).

- Make a detailed business case for the migration, including the expected cost (IDA, 2003).

- Cost is (one of) the most important reasons for migrating to open source software (Zuliani & Succi, 2004a; Brink et al., 2006; COSPA, 2005; Fitzgerald & Kenny, 2003; Waring & Maddocks, 2005).

- The migration will however imply important switching costs (e.g., training and migration), which may become a barrier (Drozdik et al., 2005; Kovács et al., 2004b, 2004a; Waring & Maddocks, 2005; IDA, 2003).

- Consequently, the real cost savings (if present at all) are difficult to quantify (Drozdik et al., 2005; Russo et al., 2003; Wichmann, 2002; COSPA, 2005).

- Other factors such as quality and productivity may also be important (Zuliani & Succi, 2004a).

(b) Pilot Study

- The use of a pilot study is recommended (IDA, 2003; Brink et al., 2006).

2. Migrating towards Open Source Software

(a) Pace of Migration

- A "big bang" approach should be avoided since it increases the risk of the migration (IDA, 2003).

- When migrating to a fully open source desktop on Linux, users should first be migrated to open source desktop software such as OpenOffice.org and Mozilla on MS Windows. In a second phase, MS Windows can be replaced by Linux (COSPA, 2005; Drozdik et al., 2005).

- When replacing MS Office by OpenOffice.org, there should be a transition phase in which users have access to both office suites to lower user resistance (Zuliani & Succi, 2004b; Zuliani & Succi, 2004a; COSPA, 2005). On the other hand, users are then likely to expect the same behavior from OpenOffice.org as from MS Office (Drozdik et al., 2005).

(b) Top Management Support

- Top management support has been found to be critical during deployment (Brink et al., 2006; Fitzgerald & Kenny, 2003).

(c) Attitude of End Users

- It is important to create a positive attitude with end users before the migration, since personnel resistance is the most important issue in the migration (Drozdik et al., 2005; Rossi et al., 2005; Zuliani & Succi, 2004b, 2004a).

- Personnel may perceive the transition negatively if they are satisfied with the current application that they may be using at home as well (Russo et al., 2003; COSPA, 2005).

- It is important to consult and communicate with users in order to minimize discomfort for users (Drozdik et al., 2005; IDA, 2003).

- Users may fear becoming "deskilled" in using a non-industry standard, thereby decreasing their value on the labor market (IDA, 2003; Fitzgerald & Kenny, 2003).

---

continued on the following page

Table 3. continued

---

3. Training

- Although users with a good knowledge of MS Office tend to experience few problems with the transition to OpenOffice.org (Kovács et al., 2004b; Kovács et al., 2004a; Russo et al., 2003), it is important that employees receive proper training to improve user acceptance (Rossi et al., 2005; Brink et al., 2006; IDA, 2003; COSPA, 2005; Kovács et al., 2004b, 2004a).

- Training is also important because some functions in OpenOffice.org behave differently than in MS Office or have different names (Jashari & Stojanovski, 2006; COSPA, 2005; Drozdik et al., 2005).

- Training should immediately precede (or follow) the migration, so that users can start practicing their skills in OpenOffice.org (Zuliani & Succi, 2004b; Zuliani & Succi, 2004a; COSPA, 2005).

- Training should focus on generic capabilities in office productivity, i.e., functionalities that are used each day (Russo et al., 2003; COSPA, 2005).

- The training approach can vary, e.g., face-to-face training, interactive tutorials over the Intranet and seminars (Russo et al., 2003; Brink et al., 2006).

4. Support

- A lack of external support for assisting in the migration can be a barrier, if the required knowledge is not available in-house (Kovács et al., 2004b; Kovács et al., 2004a; Jashari & Stojanovski, 2006).

- It is important that users have access to several kinds of support, e.g., knowledge bases containing FAQs, guides, and manuals; access to telephone and e-mail support; and access to a contact person (e.g., a product champion) in case of questions (Zuliani & Succi, 2004b; Zuliani & Succi, 2004a; Brink et al., 2006; IDA, 2003).

5. Document Conversion

- The conversion of OpenOffice.org documents from and to MS Office format does not cause many issues in most cases, although some incompatibilities may occur (e.g., wrong image size or margin settings) (COSPA, 2005; Drozdik et al., 2005; Russo et al., 2003).

- However, when using complex documents with precise formatting or MS Office macros, conversion may become a very labor-intensive task, and conversion incompatibilities may arise, especially if these files are frequently converted from MS Office to OpenOffice.org format and vice versa (COSPA, 2005; Zuliani & Succi, 2004b, 2004a; Drozdik et al., 2005).

6. Functionality

- Perceptions towards the functionality offered by OpenOffice.org tend to vary. Some users report that OpenOffice.org offers the same functionality as MS Office (Rossi et al., 2005; COSPA, 2005), while other users complain about missing features (Rossi et al., 2005).

- In general, the functionalities of OpenOffice.org seem more than adequate for daily use, and a migration is possible with no or few problems (Zuliani & Succi, 2004b; Rossi et al., 2005; COSPA, 2005).

- It should be noted, however, that depending on the use of OpenOffice.org, specific issues may arise (e.g., mail merge feature, and differences in hard and soft line breaks) (Zuliani & Succi, 2004a; COSPA, 2005).

- Interoperability and compatibility with existing systems (e.g., databases and desktop applications) can be a problem in certain situations (Kovács et al., 2004b; Kovács et al., 2004a; Zuliani & Succi, 2004a).

---

years earlier, as reported by the first FLOSS study (Wichmann, 2002). This study showed that only 6.9% of PAs and businesses used open source software on the desktop[12] (Wichmann, 2002).

Recommendations from Previous Literature

The studies that are described in the previous section, have provided more insight into the migration process that was followed by a number of PAs. Potential adopters of open source software can draw important lessons from these studies to avoid running into the same problems that previous migrations already handled effectively. Unfortunately, given the exploratory nature of this research, the literature on this topic is rather fragmented. Few attempts have been made to integrate these results, although the COSPA and tOSSad projects are currently working towards this. Hence, our aim is to integrate the results from the currently available empirical literature, to provide a comprehensive overview of the recommendations and "*lessons learned*" from previous migrations.

Although ideally we would like to present a number of guidelines or best practices, we feel that this is currently not yet feasible because of a number of reasons. First, this type of research is still exploratory, and there are still many variations in the adoption processes followed and in the context in which the migration takes place, leading to a more or less—depending on the situation—successful migration. Second, and more importantly, the studies that are currently available are not likely to be representative for all PAs. Studies in this domain tend not to report how the cases in their sample were chosen. Hence, the possibility cannot be excluded that the cases were selected out of practical considerations (e.g., having access to the site), rather than based on their theoretical relevance. Consequently, the lessons learned that are presented here should be considered preliminary, and the results of future studies should be contrasted with this set of recommendations.

Based upon previous studies, we derived a set of recommendations and lessons learned in a number of different areas. This set was derived as follows: first, the literature was searched for studies on the adoption of open source desktop software by PAs. We restricted our search to empirical studies in academic literature, discussing the experiences and results from migrations in PAs. We also chose to include the *IDA Open Source Migration Guidelines* (IDA, 2003) in the literature study. Although this is not an academic study, it was one of the first studies to provide recommendations on how PAs should migrate towards open source software. It has been reported, however, that these guidelines are not very often used in practice (Zuliani & Succi, 2004a). The studies included in this literature study are listed in Table 2. Next, the studies were analyzed by one of the authors and any lessons learned or recommendations were coded in the text. The initial coding categories were based on several aspects related to a migration process that are commonly known to be relevant (e.g., training, analysis, and document conversion). Coding was flexible and opportunistic, adding new categories when they were encountered. This initial list of lessons learned was subsequently reviewed, and data were further integrated as necessary. The revised list was then reviewed by a second author and some additions were applied as required. The result of this analysis is shown in Table 3. In total, six different areas regarding the transition were identified, namely *analysis and preparation, migration, training, support, document conversion,* and *functionality.* Each area contains a number of recommendations and lessons learned from previous research.

Case Study in the Brussels PA

In this section, we will present the results of a case study on the migration of the Government of the Brussels-Capital Region in Belgium. The Brussels-Capital Region consists of the 19 municipalities of Brussels. The Government of the Brussels-Capital Region consists of eight Ministries, each having its own cabinet. The *Brussels Regional Informatics Center (BRIC)*, responsible for the promotion and assistance of information technology (IT) within the Government of the Brussels-Capital Region, was responsible for organizing the transition towards OpenOffice.org. Our aim is to compare the approach taken by BRIC to the list of lessons learned that was compiled in the previous section. Due to space limitations, we will focus exclusively on the areas present in Table 3.

Methodology

A descriptive case study approach was used to study the transition towards OpenOffice.org at the Government of the Brussels-Capital Region. The case study approach allowed us to study the phenomenon in its real-life context (Benbasat, Goldstein, & Mead, 1987; Yin, 2003). An embedded case study design was used in order to investigate the migration towards OpenOffice.org at BRIC as well as at the ministerial cabinets of the Brussels-Capital Region. Using the key informant method, we selected two informants within BRIC, since the use of a single informant may lead to unreliable results (Benbasat et al., 1987; Phillips, 1981). Our informants were the director of the IT department and the project leader who was assigned to the OpenOffice.org project. Both informants were highly involved in the migration towards OpenOffice.org, and were responsible for planning and coordinating the migration, developing documentation, designing the training sessions, and conducting user evaluations.

The primary mode of data collection consisted of two face-to-face interviews which were conducted by a two-person team. During the first interview, important background information on the transition was gathered. Based on this information, the case study protocol was completed, leading to the generation of a detailed set of questions. During the second interview, detailed information on the migration was gathered from our informants. This interview was digitally recorded for future reference. One researcher of the team was primarily responsible for posing the interview questions, while the other was responsible for taking notes and supplementing the interview with additional questions. Using different roles for each researcher also allowed us to view the case from two perspectives and to compare the impressions of both researchers afterwards (Eisenhardt, 1989; Yin, 2003). Additional sources of evidence were internal documents of BRIC, legislative texts, and secondary information such as press releases. Extensive follow-up questions on the interview and recent developments took place via e-mail. A draft copy of the case study report, as well as a draft of this chapter, was reviewed by our informants to increase the validity of our findings.

Findings

Analysis and Preparation

Planning

The migration to OpenOffice.org in the Brussels PA was driven by two political decisions. First, a resolution was voted in which the use of open standards and open source software was encouraged by the Brussels-Capital Region in order to facilitate communication with its citizens. Consequently, BRIC was required to consider at least one open source alternative in each new project. Second, following this resolution, the coalition agreement of the Brussels-Capital Region in 2004 also stated that the use of open standards and open source software would be encouraged within the Brussels-Capital Region. Based upon this coalition agreement, the Government of the Brussels-Capital Region decided that open source office software would be used by the ministerial cabinets of the Brussels-Capital Region. OpenOffice.org was, however, not mentioned by name at that time. The migration involved installing OpenOffice.org 1.1 on 400 workstations running on MS Windows XP. In addition, four out of eight servers of the ministerial cabinets were migrated from MS Windows to Linux.

Preceding the migration, no formal TCO analysis was carried out. Although the main argument for migrating towards OpenOffice.org was to facilitate communication with citizens, cost savings realized by the migration were also stressed, especially in public announcements. License costs were cut back by 185,000 EUR during the first year and 15,000 EUR in the subsequent years (when a limited number of remaining workstations will be migrated). On the other hand, our informants confirmed that significant hidden costs of training and support occurred. Unfortunately, it was not possible to accurately quantify these hidden costs.

Pilot Study

To verify the feasibility of a transition from MS Office to OpenOffice.org, BRIC performed a pilot project in March 2004. This pilot consisted of migrating the workstations of BRIC personnel to OpenOffice.org. The outcome of the pilot project confirmed the feasibility of migrating the ministerial cabinets. As soon as OpenOffice.org 2.0 was available, BRIC again executed an internal pilot study. In contrast to the prior pilot, the aim was not only to prepare the upgrade towards OpenOffice.org 2.0, but also to enable BRIC to market the new version towards end users. They therefore invited the key users of each cabinet (i.e., the cabinet clerk and the local IT responsible) to participate in the pilot. Based on these experiences and significant improvements of OpenOffice.org 2.0 in comparison with version 1.1, it was decided in December 2005 to upgrade the ministerial cabinets to OpenOffice.org 2.0.

Migrating Towards Open Source Software

Pace of Migration

As a new Government of the Brussels-Capital Region is elected every five years, the computer equipment of the ministerial cabinets is updated simultaneously. Because of efficiency reasons, and to minimize discomfort for end users, BRIC decided to have the migration towards OpenOffice.org 1.1 coincide with the replacement of the PCs. When the user's workstation was replaced, only OpenOffice.org was installed and MS Office was no longer available to the user, except in a limited number of cases in which the user required advanced functionality of MS Excel or MS Access (e.g., in the finance department). Concurrently, the default data format for internal communication changed from MS Office to OpenOffice.org format. This change did not pose any insurmountable problems thanks to the import/export filters of OpenOffice.org. These filters enabled opening and saving previously existing MS Office documents, or MS Office documents that are sent from and to external users.

The decision to use a "big bang" approach was motivated by a number of reasons. First, BRIC judged that it would be more manageable to instantaneously switch to OpenOffice.org without temporarily installing MS Office. Furthermore, it was expected that users with an initial negative attitude towards OpenOffice.org would continue using MS Office as long as it was still available on their workstation. Finally, by changing the default document format to OpenOffice.org, BRIC wanted to encourage staff members to use OpenOffice.org.

Top Management Support

Top management support has frequently been shown to be positively related to the acquisition and successful implementation of new technologies (see e.g., DeLone, 1988; Rai & Patnayakuni, 1996). As already mentioned, the migration to OpenOffice.org was mandatory since it was a political decision. Therefore, top management (i.e., the Government of the Brussels-Capital Region) was formally supporting the migration. Given that the migration was mandated by law, BRIC was endowed with a powerful mandate while performing the migration. This helped in countering possible user resistance. It should be noted that some cabinet clerks were not in favor of the migration towards OpenOffice.org. It is likely that they have had an impact on the attitudes of end users. Unfortunately, it was not possible to empirically confirm this hypothesis.

Attitude of End Users

Within the ministerial cabinets, it could be observed that users who showed an initial negative attitude towards OpenOffice.org, remained rather skeptical. This fact confirms the importance of training and providing users with adequate information before migration takes place. The fear of becoming "deskilled" (Fitzgerald & Kenny, 2003) did not seem to have occurred. According to our informants, end users articulated no concerns about moving away from the industry standard, possibly reducing their value in the job market.

Training

According to the recommendations from previous literature, users should be able to start practicing with OpenOffice.org immediately after training. However, this was not always possible because of the narrow time frame in which the migration took place. Nevertheless, all users obtained their training within one week before or after their workstation was migrated.

The training consisted of a voluntary training course in the offices of BRIC and a CD-ROM. During the course, the basic functionality of OpenOffice.org Writer and Calc was explained. The first sessions of this course were intended for the key users of the different cabinets to enable them to provide first-line support to their users. Representatives of two out of eight ministerial cabinets found the training too basic, and did not encourage members of their cabinet to attend the training sessions. Afterwards, a survey among end users indicated that users who did not attend the course reported more problems in using OpenOffice.org.

It was observed that people tried to work in the same manner as they were used to in MS Office. This is however not always possible because some functions in OpenOffice.org are fundamentally different from their MS Office counterpart. Therefore, short personal demonstrations were organized during which BRIC personnel illustrated the procedure to be used in OpenOffice.org.

Although the general training session proved to be useful, BRIC noticed that additional, focused training sessions would be required following the upgrade to OpenOffice.org 2.0, due to functional differences with OpenOffice.org 1.1. These sessions will be organized as short workshops: each session will focus on a particular functionality that different groups of end users require (e.g., using the Mail Merge feature or working with document templates). The workshops will also be more practically oriented than the initial training sessions.

Support

Three important sources of support are available to end users in the ministerial cabinets. First-line support for common problems can be provided by the key users in each cabinet. Second, BRIC itself provides end user support for OpenOffice.org. Third, during the pilot study and training, BRIC employees have built an extensive knowledge base on OpenOffice.org. This knowledge base—which is regularly updated by new questions formulated by end users—is sufficient to solve most problems.

Part of the knowledge base consists of a CD-ROM containing a manual, a FAQ list and the OpenOffice.org installation files. This CD-ROM was handed out to the participants of the training session. The manual and FAQ list were based on the documentation provided by the OpenOffice.org communities. However, since the Brussels-Capital Region is bilingual (with Dutch and French being the official languages), documentation had to be sourced from two different OpenOffice.org localization communities. It then became apparent that significant differences in quality between these online communities exist. While the French community is very vivid and provides much information, the Dutch counterpart does not produce the same quantity of documentation. This did lead to some difficulties, since BRIC is required to provide (equivalent) training material in both languages. Another indication of the difference between the French and Dutch community are delays in the release of upgrades. For both OpenOffice.org 1.1 and 2.0, the French localization was available much earlier than the Dutch version. This has been criticized by some end users as a disadvantage of working with open source software.

Document Conversion

In general, few problems concerning document conversion between MS Office and OpenOffice.org were reported. However, in three specific cases, some issues did

Figure 1. Formatting of legislative texts in MS Office vs. OpenOffice.org

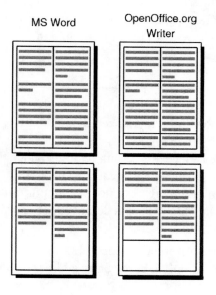

occur. First, difficulties arose when documents with extensive and complex layout were exchanged with external parties. When these documents were converted several times back and forth between OpenOffice.org and MS Office format, incompatibilities in layout became unmanageable. As a result, it was decided to exchange documents without formatting as long as they needed revision. Once the final version of the document had been approved, the formatting was done either in OpenOffice.org or in MS Office. The use of the PDF-format was also promoted for documents requiring no further revisions.

Secondly, the conversion of MS Office document templates to OpenOffice.org posed some problems, mainly because a number of incompatibilities exist between the two products, for instance, with respect to margin settings. Since the Brussels PA has a very rigorous style guide, these templates had to be migrated very accurately. As a result, BRIC only recently finalized the OpenOffice.org templates.

A third issue, concerning the editing of legislative texts, was again caused by very specific format prescriptions. Since the Brussels-Capital Region is bilingual, these texts have to be published in two columns, one for each language. Moreover, each paragraph has to start on the same height as the corresponding paragraph in the other language. Given that French paragraphs are somewhat lengthier than the Dutch equivalents, some adjustments in vertical spacing between paragraphs must be made (see Figure 1). In MS Word this layout was realized by using a table with two columns and one row. OpenOffice.org 1.1, however, did not support multi-page table cells. Hence, the layout had to be changed by putting each paragraph in a different cell. When a cell does not fit on a single page, it is moved automatically to the next page. In OpenOffice.org 2.0 this workaround has become obsolete, since it is possible to use multi-page rows in the new version[13].

Functionality

Regarding functionality, several remarks can be made. First, a number of end users did not consider OpenOffice.org a fully fledged alternative to MS Office. This was mainly caused by the absence of certain features which are present in MS Office. The fact that MS Excel and MS Access are still used by a limited number of users, confirms that OpenOffice.org does not yet support all advanced features of MS Office. On the other hand, most users do not experience any problems in their daily tasks.

Furthermore, users reported problems concerning the Mail Merge feature. When using this feature to create a mailing based on an address list in a database, OpenOffice.org creates a new document for each addressee. To solve this issue, a script was developed to merge the separate files into one document. In OpenOffice.org 2.0, however, this problem does not occur anymore because the user can choose whether to generate one large file or to generate separate documents. It is therefore hoped that the improved functionality of OpenOffice.org 2.0 will enhance the perceived quality of OpenOffice.org by end users. Nevertheless, a number of deficiencies are still reported by users working with OpenOffice.org 2.0. Although workarounds are possible for most of these problems, these defects still have a negative impact on the general perception of OpenOffice.org.

The use of data sources by OpenOffice.org is considered to be a mixed advantage and disadvantage. In contrast to MS Office, OpenOffice.org allows a document to access more than one data source. The implementation of these data sources is, however, significantly different from MS Office: while in MS Office the data source is included in the document itself, OpenOffice.org stores it in the user profile at the user's workstation. Thus, when exchanging the document between users, the data source is lost. To solve this problem, BRIC wrote a script to ensure that each user has access to the commonly used data sources.

The perceived usability of OpenOffice.org 1.1 is also reported to be inferior compared to MS Office. The most often heard critique is that the look and feel of OpenOffice.org 1.1 feels outdated. Again, OpenOffice.org 2.0 could provide a solution because its look and feel has been updated considerably to resemble that of MS Office.

Discussion

In general, the findings in this case study are consistent with the experiences and recommendations from previous literature. The case study illustrated that a migration towards OpenOffice.org was possible within the Government of the Brussels-Capital Region. Nevertheless, the case study also confirms that there are a number of important issues that should be paid attention to when planning and performing the migration.

Following the recommendations from previous literature, it can be observed that the documentation available and training given to end users are very important. With respect to training, the case study showed that in a first phase, the training can focus on general office suite capabilities applied to OpenOffice.org. Therefore, it serves as a revision course of skills that are also applicable to MS Office. This is consistent with recommendations made in previous literature (see, e.g., COSPA, 2005; Russo, Zuliani, & Succi, 2003). In a second phase, however, it seemed useful to offer short, focused training sessions on particular tasks in working with OpenOffice.org. Both training and documentation can positively influence the attitude of end users. While the resistance of end users is frequently a major problem (Rossi, et al., 2005; Zuliani & Succi, 2004b), we feel that in this case the acceptance was facilitated not only by strong top management support, but also by the fact that the adoption of OpenOffice.org was mandated by law, giving end users no choice but to conform with the transition. In terms of interoperability, the document conversion facilities of OpenOffice.org seemed to be sufficient in most cases (COSPA, 2005; Russo et al., 2003). However, advanced use of OpenOffice.org, or specific applications (e.g., bilingual legislative texts) may lead to certain issues (COSPA, 2005; Zuliani & Succi, 2004a). Conducting a pilot test may help in identifying potential issues that arise during implementation within the specific context in which the migration will take place. Although the decision to start using OpenOffice.org was primarily politically motivated, there was an emphasis on the cost savings realized by the transition. It must be noted, however, that considerable effort was invested in developing appropriate training, documentation, and document templates.

One important difference with previous studies is that the "big bang" approach taken by BRIC did seem successful. Previous studies have recommended that a transition phase is planned in which MS Office and OpenOffice.org are installed concurrently on the users' workstation (COSPA, 2005; Zuliani & Succi, 2004a, 2004b). Based on the present study, we feel that there are situations in which MS Office can be replaced immediately with OpenOffice.org. While the transition period enables users to continue using MS Office in case they experience problems with OpenOffice.org, the immediate migration towards OpenOffice.org forces users to start using (and learning) OpenOffice.org. By having to rely solely on OpenOffice.org to complete office productivity tasks, users might become familiar more quickly with the functionality of OpenOffice.org, shortening the learning period.

It also encourages users to start adopting the preferred OpenOffice.org methods, instead of trying to work with OpenOffice.org in the same way as with MS Office (Drozdik, et al., 2005). Therefore, we feel a "big bang" migration from MS Office to OpenOffice.org is feasible in certain situations. On the other hand, if a PA would decide to replace the MS Windows desktop with Linux desktops, it seems better to use a phased approach, as recommended in previous studies (COSPA, 2005; Drozdik, et al., 2005) (see also Table 3).

Implications

This study has a number of implications for both practice as well as research.

Implications for Practice

Several PAs have already decided to migrate towards open source software based solutions or have already completed their migration. It is likely that, thanks to the encouragement of open standards and open source software by the European Commission, many other PAs will also decide to migrate to open source software. The results of currently ongoing research projects such as COSPA, FLOSSPOLS, and tOSSad will probably further help PAs to decide on the possible risks and advantages. The lessons learned identified in this chapter should serve as a first step towards developing research-based best practices for the adoption of open source software on the desktop. The presented overview can be of help to PAs contemplating a migration. It will assist in identifying possible barriers or issues that should be anticipated in the project to avoid difficulties during implementation. As a result, PAs can learn from previous efforts of early adopters. We believe that by promoting the sharing of experiences of such migrations, future migrations can be handled more effectively (both in time and financially).

Based on available research, we believe that a migration towards open source software on the desktop is possible, but is contingent on the specific environment in which the migration will take place. If the PA is using advanced features of MS Office (e.g., the extensive use of macros), or when interoperability with existing systems and applications is important (e.g., database servers, or proprietary tools written for MS Office), it will be more difficult to migrate. Although workarounds can be devised, this will imply additional costs, and productivity may be negatively affected. We believe, however, that as more organizations and PAs decide to start using open standards and open source software, software vendors will be more likely to provide interoperability with open source solutions[14].

When the use of MS Office is mainly restricted to basic functionality—which seems to be the case in most PAs currently studied in literature—the functionality of OpenOffice.org should suffice in most cases. Nevertheless, PAs should consider the issues raised in this chapter, and sufficient attention should be paid to areas such as planning, training, and support. Moreover, PAs should be aware that specific requirements in their environments may lead to a number of issues that were not previously encountered. For example, due to the bilingualism in the Brussels PA, the implementation of tables in OpenOffice.org resulted in changing the way documents were formatted. This implied altering the daily work habits of end users, which may prove to be very difficult. Hence, having top management support—and possibly a product champion—may be required to ensure that new working practices are adopted. Furthermore, the use of a pilot study may help in encountering issues in the daily work habits of employees that need to be resolved before migrating.

Our results may also be relevant to the open source community. Results from previous studies, as well as the current study, indicate that one of the main drivers for a migration towards open source software is cost reduction. This is consistent with literature on the organizational adoption of open source software, which has shown that the lower or non-existing license costs of open source software is a major reason influencing the adoption decision (see e.g., Dedrick & West, 2003; Ven & Verelst, 2006). This is in contrast with the attitude of the open source community that tries to downplay the cost advantages, and tends to emphasize other advantages such as having access to the source code of the application, and being allowed to modify it. Studies in this field may offer the open source community a better insight into the perceptions of organizations and PAs, since it has been noted that the open source community in general has limited access to the opinion of its customers (Fitzgerald, 2005). A more profound insight into the main motivations of adopters of open source software may lead the open source community (and especially the open source vendors) to emphasize different advantages of open source software. Similarly, open source communities (especially projects such as OpenOffice.org and Mozilla) may also try to actively solicit feedback from PAs in order to include missing features and further improve functionality.

Implications for Research

This study has a number of limitations. First, given the fact that the migration of PAs towards open source software on the desktop is a new phenomenon, relatively little research has been devoted to studying the migration process. Consequently, our lessons learned are based on the limited amount of studies that were available at the time of writing. Furthermore, it is difficult to assess the representativeness of these studies. Second, we compared the lessons learned with a single case study in the Brussels PA. Hence, the external validity of this study cannot be ascertained, and we therefore cannot generalize our results to all PAs. Future research can therefore study additional migrations that can be compared to the issues raised in this chapter. By incorporating additional findings from different contexts, a set of research-based best practices for the adoption of open source software on the desktop may be deducted. These may further help clarifying the factors that influence a successful migration. This study should be considered a first step in this direction.

Another interesting avenue for further research is to investigate to which degree the migration process in PAs is different from the process in enterprises. For example, PAs may be better suited to motivate—or force—their users to make the migration succeed. In the case of the Brussels PA, the migration towards OpenOffice.org was mandated by law, making it very difficult to resist this change. It is likely that enterprises are not capable of supporting the migration in such a strong way. Second, it is possible that enterprises generally make more advanced use of MS Office, or use it in a more complex environment, making the migration more difficult.

Conclusion

There has been an increased interest in open source software by PAs in the past few years. Some authors expect that this trend will continue in the following years, making PAs a driving force behind open source software in Europe (González-Barahona & Robles, 2005). Lately, much attention has been given to migrations towards open source software on the desktop. Given the apparent success of several attempts in this direction, we expect that PAs will continue to consider these migrations in the future. However, given the fact that a migration towards open source software on the desktop is quite disruptive for end users, it is important that guidelines are available. The recommendations deducted from previous research that are listed in this chapter can be considered a first step in this direction. By comparing the lessons learned from previous migrations to a case in the Brussels PA, it was noted that, in general, there was a good match between the literature and the current case. Due to the specific context of the case, we were able to highlight a few issues that may be of interest for future adopters. Furthermore, we feel that it is important to conduct additional case studies on the migration towards open source software and compare the results to the recommendations described in this chapter. This way, the recommendations will be based on a substantial body of knowledge, leading to a set of best practices for the migration towards open source software on the desktop.

Trademark Use

Microsoft Office and Microsoft Windows are registered trademarks of Microsoft Corporation. OpenOffice.org is a registered trademark of Team OpenOffice.org e.V.

References

Applewhite, A. (2003). Should governments go open source? *IEEE Software, 20*(4), 88-91.

Benbasat, I., Goldstein, D. K., & Mead, M. (1987). The case research strategy in studies of information systems. *MIS Quarterly, 11*(3), 368-386.

Bretschneider, S., & Wittmer, D. (1993). Organizational adoption of microcomputer technology: The role of sector. *Information Systems Research, 4*(1), 88-108.

Brink, D., Roos, L., Weller, J., & van Belle, J.-P. (2006). Critical success factors for migrating to OSS-on-the-desktop: Common themes across three South African case studies. In E. Damiani, B. Fitzgerald, W. Scacchi, M. Scotto, & G. Succi (Eds.), *IFIP International Federation for Information Processing* (Vol. 203, pp. 287-293, open source system). Boston: Springer.

Canonico, P. (2005). Deploying open source applications within the public sector domain: Preliminary findings on potential organisational benefits and drawbacks. In G. Mangia & R. Mohr (Eds.), *Proceedings of the German-Italian Workshop on Information Systems (GIWIS 2005)* (pp. 85-92).

COSPA Project. (2005). *Work package 4, deliverable 4.3—Experience report on the implementation of OS applications in the partner PAs.* Retrieved July 4, 2006, from http://www.cospa-project.org/ download_access.php?file=D4.3-ExperienceReportOnTheImplementationOfOS.pdf

Dedrick, J., & West, J. (2003). Why firms adopt open source platforms: A grounded theory of innovation and standards adoption. In J. L. King & K. Lyytinen (Eds.), *Proceedings of the Workshop on Standard Making: A Critical Research Frontier for Information Systems* (pp. 236-257).

DeLone, W. H. (1988). Determinants of success for computer usage in small business. *MIS Quarterly, 12*(1), 50-61.

Drozdik, S., Kovács, G. L., & Kochis, P. Z. (2005). Risk assessment of an open source migration project. In M. Scotto & G. Succi (Eds.), *Proceedings of the First International Conference on Open Source Systems* (pp. 246-249).

Eisenhardt, K. M. (1989). Building theories from case study research. *Academy of Management Review*, *14*(4), 532-550.

Fitzgerald, B. (2005). Has open source software a future? In J. Feller, B. Fitzgerald, S. Hissam, & K. Lakhani (Eds.), *Perspectives on free and open source software* (pp. 93-106). Cambridge, MA: MIT Press.

Fitzgerald, B., & Kenny, T. (2003). Open source software in the trenches: Lessons from a large scale implementation. In S. T. March, A. Massey, & J. I. DeGross (Eds.), *Proceedings of 24th International Conference on Information Systems (ICIS)* (pp. 316-326).

Ghosh, R., & Glott, R. (2005). *Free/libre and open source software: Policy support (FLOSSPOLS)—deliverable D3: Results and policy paper from survey of government authorities.* Retrieved August 5, 2005, from http://flosspols. org/deliverables/ FLOSSPOLS-D03 (MERIT, University of Maastricht)

González-Barahona, J. M., & Robles, G. (2005). Libre software in Europe. In C. Dibona, D. Cooper & M. Stone (Eds.), *Open sources 2.0* (pp. 161-188). Sebastopol, California: O'Reilly.

IDA—Interchange of Data between Administrations. (2003). *The IDA open source migration guidelines.* Retrieved October 23, 2003, from http://europa.eu.int/ idabc/servlets/Doc?id=1983

Jashari, B., & Stojanovski, F. (2006). Challenges and obstacles: Usage of free and open source software in local government in Macedonia. In B. Özel, C. B. Çilingir, & K. Erkan (Eds.), *Proceedings of tOSSad OSS2006 Workshop: Towards Open Source Software Adoption: Educational, Public, Legal, and Usability Practices* (pp. 49-55). Kocaeli, Turkey: Tübitak.

Kovács, G. L., Drozdik, S., Zuliani, P., & Succi, G. (2004a). Open source software and open data standards in public administration. In *Proceedings of the IEEE International Conf. on Computational Cybernetics (ICCC2004)* (pp. 421-428).

Kovács, G. L., Drozdik, S., Zuliani, P., & Succi, G. (2004b). Open source software for the public administration. In *Proceedings of the 6th Computer Science and Information Technologies (CSIT)* (pp. 1-8).

Lee, J.-A. (2006). Government policy toward open source software: The puzzles of neutrality and competition. *Knowledge, Technology, & Policy*, *18*(4), 113-141.

Lundell, B., Lings, B., & Lindqvist, E. (2006). Perceptions and uptake of open source in Swedish organisations. In E. Damiani, B. Fitzgerald, W. Scacchi, M. Scotto & G. Succi (Eds.) *IFIP International Federation for Information Processing* (Vol. 203, pp. 155-163, open source systems). Boston: Springer.

Moon, M. J., & Bretschneider, S. (2002). Does the perception of red tape constrain IT innovativeness in organizations? Unexpected results from a simultaneous equation model and implications. *Journal of Public Administration Research & Theory*, *12*(2), 273-291.

Nye, J. S. (1999). Information technology and democratic governance. In E.C. Kamarck & J. S. Nye (Eds.), *Democracy.com? Governance in a networked world*. Hollis, NH: Hollis Publishing (pp. 1-18).

Phillips, L. W. (1981). Assessing measurement error in key informant reports: A methodological note on organizational analysis in marketing. *Journal of Marketing Research, 18*(4), 395-415.

Rai, A., & Patnayakuni, R. (1996). A structural model for CASE adoption behavior. *Journal of Management Information Systems, 13*(2), 205-234.

Rossi, B., Russo, B., & Succi, G. (2006). A study on the introduction of open source software in the public administration. In E. Damiani, B. Fitzgerald, W. Scacchi, M. Scotto, & G. Succi (Eds.). *IFIP International Federation for Information Processing* (Vol. 203, pp. 165-171, open source systems). Boston, MA: Springer.

Rossi, B., Scotto, M., Sillitti, A., & Succi, G. (2005). Criteria for the non invasive transition to openoffice. In M. Scotto & G. Succi (Eds.), *Proceedings of the First International Conference on Open Source Systems* (pp. 250-253).

Russo, B., Zuliani, P., & Succi, G. (2003). Toward an empirical assessment of the benefits of open source software. In J. Feller, B. Fitzgerald, S. A. Hissam & K. Lakhani (Eds.), *Taking stock of the bazaar: Proceedings of the Third ICSE Workshop on Open Source Software Engineering* (pp. 117-120).

tOSSad. (2006). *F/OSS National Programme Start-Up Roadmap Report*. Retrieved June 28, 2006, from http://tossad.org/tossad/publications/ f_oss_national_programme_start_up_report_1

Vaca, A. (2005). Extremadura and the revolution of free software. In M. Wynants & J. Cornelis (Eds.), *How open is the future? Economic, social & cultural scenarios inspired by free and open source software* (pp. 167-197). Brussels, Belgium: VUB Brussels University Press.

Ven, K., & Verelst, J. (2006). The organizational adoption of open source server software by Belgian organizations. In E. Damiani, B. Fitzgerald, W. Scacchi, M. Scotto, & G. Succi (Eds.), *IFIP International Federation for Information Processing* (pp. 111-122). Boston, MA: Springer.

Waring, T., & Maddocks, P. (2005). Open source software implementation in the UK public sector: Evidence from the field and implications for the future. *International Journal of Information Management, 25*(5), 411-428.

Wichmann, T. (2002). *FLOSS final report—Part 1: Use of open source software in firms and public institutions—Evidence from Germany, Sweden and UK*. Retrieved September 8, 2003, from http://www.infonomics.nl/FLOSS/report/ reportPart1_use_oss_in_firms_and_public_institutions.htm

Yin, R. K. (2003). *Case study research: Design and methods* (3rd ed.). Newbury Park, California: Sage Publications.

Zuliani, P., & Succi, G. (2004a). An experience of transition to open source software in local authorities. In *Proceedings of E-Challenges on Software Engineering.*

Zuliani, P., & Succi, G. (2004b). Migrating public administrations to open source software. In P. Isaías, P. Kommers, & M. McPherson (Eds.). *Proceedings of E-Society 2004 IADIS International Conference* (pp. 829-832). IADIS Press.

Endnotes

[1]   For a thorough discussion of the concerns involving the use of open source software by PAs, see Lee (2006).

[2]   Although this group is currently inactive, some information on it can still be retrieved at http://eu.conecta.it

[3]   http://europa.eu.int/idabc

[4]   See http://europa.eu.int/information_society/eeurope/2005/all_about/'action_plan/index_en.htm

[5]   http://europa.eu.int/idabc/en/chapter/452

[6]   http://www.cospa-project.org

[7]   http://www.flosspols.org

[8]   http://www.calibre.ie

[9]   http://www.tossad.org

[10] See IDABC "open source case studies" (http://ec.europa.eu/idabc/en/chapter/470) and news (http://ec.europa.eu/idabc/en/chapter/491).

[11] http://www.muenchen.de/Rathaus/referate/dir/limux/89256/

[12] It must be noted that the results of these two studies cannot be directly compared. The first FLOSS study included commercial businesses as well as PAs, and included subjects from only 3 European countries.

[13] The workaround is however still used, since it is a better approach for formatting these documents.

[14] Actually, at the time of writing, Microsoft has announced its support to an independent open source project, developing an ODF plug-in (Open Document Format) for MS Word 2007 (http://odf-converter.sourceforge.net). Coincidently, this announcement was issued a number of days after the Belgian government announced that it would start using the ODF-format exclusively from September 2008. Other governments (e.g., the state of Massachusetts) have announced similar initiatives, or are studying them.

Chapter X

Toward a GNU/Linux Distribution for Corporate Environments

Francesco Di Cerbo, Free University of Bolzano-Bozen, Italy

Marco Scotto, Free University of Bolzano-Bozen, Italy

Alberto Sillitti, Free University of Bolzano-Bozen, Italy

Giancarlo Succi, Free University of Bolzano-Bozen, Italy

Tullio Vernazza, University of Genoa, Italy

Abstract

The introduction of a GNU/Linux-based desktop system in a large company is often problematic. In literature, several crucial issues represent such a burden, which is often cost effective for SMEs and public administrations. Some of these are technical issues; the others are related to the training costs for the employees. Mainly, the technical obstacles are represented by different hardware configurations that might require several adhoc activities to adapt a standard GNU/Linux distribution to the specific environment, including the applications profile of the company. On the other hand, to lower the learning curve of employers, we decided to work toward adopting some GNU/Linux live distributions features. In this way, we added to our project specific functionalities, which provide new and interesting capabilities to our community of users, such as self-configuration and better usability, without losing compatibility with original distributions, which is too costly in a professional scenario for its greater maintenance cost. DSS[1] (debased scripts set) tries to address the issues we mentioned above. It is a next-generation hybrid (both live and regular) distribution that includes an unmodified Debian-based GNU/Linux release and a modular-designed file system with some extended features, which we will describe in this chapter. We will also discuss the interactions with other open source communities and the positive mutual influence on DSS development process.

Introduction

The massive installation of an operating system on desktop computers in a professional scenario usually relies upon proprietary solutions, as well as the choice of adequate deployment tools. Experiences collected in the EU-funded COSPA project (see COSPA, 2003-2006) aimed to introduce open source software into PAs in Europe and showed that all PAs involved in the project had an IT infrastructure based on Microsoft Windows systems for their desktop computers.

This reliance happens because proprietary operating systems and their deployment tools are tailored and strongly rely on their capability to lower the total costs of installation and maintenance, which, first of all, consist of savings in terms of time required to implement. This is achieved thanks to the automation of many steps necessary to perform the deployment operations. Although the same facilities can be found in the GNU/Linux (Stallman, 2004) community, there is no such structured approach to the problem. Moreover, there are other issues to take into account, depending on the specific scenario: If we consider a migration from an existing IT infrastructure, the biggest problems are portability and compatibility of currently adopted software. The fulfillment of these requirements might not be guaranteed, especially when considering legacy software: This is the case of many public administrations in Italy (Assinform, 2006) in which software providers refuse to develop software for non-WIN32 environments. For these reasons, the introduction of GNU/Linux desktop systems in large companies or in public administrations is often problematic both during startup and for subsequent maintenance operations.

The different hardware configuration of target computers is one of the major technical problems with installation, since it may require several adhoc activities to adapt a standard GNU/Linux release to the specific environment. Another problem, from the users' point of view, may be represented by training/learning costs, in order to allow the staff to achieve the same productivity level they had on the previously adopted environment.

In recent years, the coming of GNU/Linux "live" distributions provided the free/open source software community with new and interesting capabilities, such as self-configuration and better usability. A "live" distribution is a particular type of GNU/Linux distribution, usually shipped within a read-only removable medium, like CD/DVD, or USB pen-drive. Live distributions cannot save any configuration information in their distribution media and may rely on several adaptation capabilities, which allow them to self-configure themselves in the system they are loaded into. We will illustrate live GNU/Linux distributions features later on in the chapter. Unfortunately, live distributions present a drawback: They are almost always derived from existing GNU/Linux distributions, but they present some relevant differences from them. Such differences make these kinds of operating systems almost useless in an ordinary professional scenario: Live distributions become forks of the original distributions and lose backward-compatibility, which means higher costs for maintenance operations, such as separate management for software upgrades and security patches adoption.

DSS (debased scripts set) tries to address the previous issues. It is a powerful GNU/Linux meta-distribution which incorporates "live" capabilities natively, based on an unmodified Debian-based (Murdock, 2004) GNU/Linux release (Ubuntu, see Ubuntu Group, 2006), including a pure "stock" kernel, that is, a standard distribution-provided Linux kernel (Rustling, 1996-1999) in binary form. DSS includes innovative hardware detection and configuration techniques, based on sound and largely adopted software (such as the hotplug daemon, as described in the Hotplug community, 2006), that is loaded during the very first boot operations. In addition, the use of a special unification file system (UnionFS, see Wright et al., 2004) along with a modular software package approach, allows DSS to deploy, in a single package, a customized company-specific release containing both the operating system and all the desired applications.

In summary, DSS is a framework that allows an easy customization of a 100% Debian-based GNU/Linux distribution, which is also at the same time a "live" one. If there is the need to perform some modifications from the standard provided version (and this is common because of an insufficient default application profile), DSS provides Debian-based tools to repackage all the modifications into an extended DSS GNU/Linux distribution. Moreover, thanks to its smart file system design, completely built by modular parts loaded at runtime, it can be easily repackaged again into a live distribution.

In this chapter, we illustrate the features of DSS, its history and connections with other free/open source software communities, and we rely on this experience to make some final considerations.

GNU/Linux Distributions: "Live" and "Regular" Approach

An ordinary GNU/Linux distribution is an aggregation of software, developed and maintained by several free/open source communities, like the GNU project (see again Stallman, 2004), which encloses a Linux kernel.

The kernel (Rustling, 1996-1999) is a necessary component because it permits the greatest portion of hardware-software interactions (through its hardware drivers), as well as the management of other running programs. However, there are other companion programs, which are necessary in several stages in running the GNU/Linux operating system. For instance, the already mentioned hotplug performs hardware events monitoring for all devices the kernel is able to manage.

Usually, these programs, as well as their configurations and the kernel itself, are stored in a hard drive partition, or, in any event a writable device, and it is very rare to modify these data and information during the startup phase. In a "regular" approach, the boot process is a sequence of operations which may be divided in two main stages: the first, which mainly performs a small setup environment, required to load the kernel into random access memory of the system, and executes it; the second, which launches all the required companion software. Both of them require their set-up configurations, which, for example, are commonly generated during an installation process.

A "live" approach relies on the concept that, as the kernel itself just needs an appropriate launching environment, it is possible to generate it dynamically at runtime, without previously existing configurations. In this way, it is possible to boot a system from a read-only device, like a CD-/DVD-ROM, and then, throughout an extensive use of system memory, to create all necessary software structures needed to boot the kernel and the companion processes. This is heavier in terms of load on system memory compared to the "regular" approach, but after the system is booted, there are no significant differences from the point of view of users, except for a possible slowdown due to intense use of the boot medium, which usually has a slower access time. Hardware configurations, as well as any parameters, are considered unknown and specific software solutions are developed to overcome these issues. Since there is almost no need for any configuration specifications, a live GNU/Linux distribution is particularly suited for un- or poorly-skilled users, who may just boot their machine and therefore use a completely working operating system without any additional effort. Usually, a live distribution is generated by a regular distribution, from which inherits software packages and their managing techniques; however, the two types differ notably in their maintenance and upgrade processes.

The drawbacks of this approach are essentially represented by the overhead, due to increased memory required, if compared to regularly installed distributions, by the frequent accesses to a medium which is usually not as fast as a hard disk, and by the strong modifications and customizations required to adapt the set-up environment. Generally speaking, maintenance and upgrade tasks performed on a live GNU/Linux distribution require more (and sometimes much more) effort than the same tasks would be on the same original (regular) distribution. The reason for such an increment in time is usually due to the tight coupling among specific versions of the software that compose a specific release of a live distribution: It may happen that modifying (upgrading or downgrading) a single component causes severe damage in the boot process. As a result, these operations require deep and accurate risk estimation, and are usually avoided by most users. Moreover, such estimation does not guarantee the feasibility of the system change which is often not acceptable for an enterprise environment.

State of the Art in Live Distributions: Knoppix[2]

Knoppix (Knopper, 2000) may be considered the pioneer of GNU/Linux live distributions, either for diffusion, as demonstrated by a large number of works based on it, or historical reasons.

However, its approach in providing a Debian GNU/Linux distribution able to boot from removable storage media, makes its use, in a professional setting, practically impossible except for data recovery or hardware testing. Its severe modifications to the standard Debian distribution, cross-combined with unstable and testing Debian applications profiles (the Debian "trees"), make the distribution and upgrade of new applications quite difficult, requiring a great effort to stabilize a new hypothetical desktop installation based on Knoppix. Moreover, "exotic" hardware suffers with Knoppix deep-kernel specificity; such specificity is a constraint due to hardware detection requirements. Uncommon, or not completely supported, hardware often comes with drivers usually not contained in standard kernels, which may even be provided with commercial licenses, and is incompatible with GPL (GNU General Public License, described in Free Software Foundation, 1991) statements and thus undeliverable, along with Debian distributions (see Debian, 2004). In these cases, the adoption of Knoppix can be a severe problem. Last, hardware detection and configuration techniques come with special boot applications (such as knoppix-autoconfig, described in Suzak et al., 2005), that require a constant maintenance process to be able to recognize new or uncommon hardware. Moreover, their approach, based on kernel-space routines, prevents successive setups (e.g., file systems configuration) from using user-space libraries and applications, which could give the user flexibility in data and device access, especially in the case of plug-and-play USB hardware. These boot applications use adhoc scripts running with maximum privileges, which may lead to security problems, particularly critical in an industrial environment. Beside these technical aspects, Knoppix introduced some interesting usability features at the time, enhancing KDE window manager (KDE e.V., 2006) to allow, for example, an easier management of devices, including removable ones, exploiting its hardware detection skills.

DSS Motivation

Our experience started when studying the introduction of a GNU/Linux distribution into a Public Administration IT infrastructure of the University of Genoa. This is a very heterogeneous scenario, with a complex user base, composed of some GNU/Linux skilled users and some less experienced ones, and a non-standardized hardware set. It was the third quarter of 2005.

This is a paradigmatic use case, which summarizes almost all the important critical features of a medium-sized IT structure.

To lower the total IT staff effort, we looked at the hardware auto-configuration skills of live distributions, starting with Knoppix. When we discovered the great maintenance problems we mentioned, which made it unsuitable for its adoption for anything different than a live or recovery system[3], we decided not to investigate further into customization tools based on Knoppix. In order to use some well-known software architectures, which would decrease the maintenance effort of our distribution in the middle term, we then moved to study the usual automatic hardware detection and configuration systems in a standard GNU/Linux environment, and we worked to introduce them in a live environment. We succeeded, as described later on, and in this way we solved a part of the problem, with the collateral benefit of the unique management of an installed GNU/Linux system, as a live one, which actually behaves regularly and without any adhoc binary program or patch to common Debian components. Then we started working on improving the overall usability, following the same approach: We looked at standard operating system event monitors and notification dispatchers inside different window managers, extending, based on an existing solution, a notification and action performer daemon able, for example, to detect and react to USB devices plugged in and out the system.

Free/open source software obtained from different communities helped us a lot to develop DSS. As they were also under development, which is common for most "successful" open source projects (Crowston, 2003), we requested some changes in the features of the software, which were gladly accepted. In this way, both projects received benefits: We simplified our development process, the others received more bug reporting and fixes due to the improved correlation rate (they "gained a client"). We did also when another community asked us for some changes and adaptations: Our project manager analyzed them and decided they were useful, so we included them into DSS and we "gained a client," a precious ally in our development process.

Our use case, at the time of writing, is still in progress, and it is not possible to provide results of this adoption to end users: From a technical point of view, we have designed and implemented the whole distribution infrastructure; we provided the community with several releases, and we are waiting for feedback. We conducted a preliminary usability test by choosing testers with different experience in the use of GNU/Linux distributions, from the unskilled to the expert level. The results of this test, however, lack statistical significance due to the small number (10) of users involved. They were indeed very useful in order to trigger our development process, giving us preliminary feedback before our first public release. There is a larger, statistically relevant migration test planned for first quarter of 2007, which would provide us with more useful results, and will be another milestone towards DSS introduction in the official IT infrastructure of University of Genoa.

DSS Main Features

DSS main features address two among the greatest issues mentioned in the GNU/Linux introduction for desktop systems into a complex IT structure. They are the plurality of configuration profiles in target desktop PCs and the distribution/maintenance of the applications. These problems may be expressed as how to cope with the complexity of managing an arbitrary number of target desktop PCs with an arbitrary number of different hardware features, minimizing human effort.

Arbitrary Configuration Profiles: Self-Configuration Techniques

Considering an arbitrary number of hardware configurations within a corporation, we decided to integrate self-configuration techniques inside DSS. In this way, the total cost of installation would be lowered by decreasing the number of required operations. As mentioned previously in the introduction, we studied the technicalities of live distribution to perform this task, then acquired and adapted them to our needs.

While developing our solution, we actually turned a "regular" GNU/Linux distribution, Ubuntu, into a "hybrid" one. In fact, while extending the usually adopted default hardware detection capabilities of Ubuntu, we adopted a special paradigm in the boot phase which could be used with almost no drawbacks either in a live or a regular distribution, actually merging the two approaches. This presents a great advantage compared to ordinary live distributions: There are no differences between DSS and its original distribution, except for a small number of scripts. This let DSS be administered and maintained as Ubuntu, which provides a lot of tools to perform those operations, and also a large community in charge of releasing updates.

DSS adopts a completely new approach to live distributions based on an "early user-space" (as described in Petullo, 2005) mode. It is a set of libraries and programs (that are available even without a running Linux kernel) which provide various functionalities required while a Linux kernel is coming up. In the forthcoming section, a technical description of this aspect will be provided.

The "early user space" mode allows DSS to use hotplug, a daemon program normally used for hardware discovery and configuration in standard non-live GNU/Linux distributions, from the first seconds of boot process. This is a great advantage, as the booting kernel relies on already detected hardware, and using its 2.6 series features, may automatically load needed kernel modules in order to use just discovered hardware, as in a regularly installed GNU/Linux system. Due to this feature, DSS does not require developing and maintaining an adhoc kernel, but it may use a stock one, exactly like any other Debian release.

To summarize, with the adoption of early user-space, except for a small set of plain scripts (very easy and simple to maintain and modify) which effectively coordinate the boot process, no adhoc component is used to bring in a Debian GNU/Linux release as a "live" distribution; moreover, the set of scripts is also completely able to fulfill hardware-automatic detection/configuration needs for kernel-supported peripherals.

Maintenance and Distribution Operations of Applications: DSS Strategies with UnionFS

In commercial operating systems, there are facilities to perform the deployment of new applications as well as maintenance for those already installed; for example, Microsoft Active Directory services with group policies[4]. The valuable benefit they provide is a centralization in the management, combined with an automation of the deployment process. In this way, it is possible to deliver automatically new software in an arbitrary broad network without any required effort on single target nodes.

DSS provides several technical solutions fit for its adoption in large-scale IT structures: an easy customization process and a centralization of IT management procedures in a whole network.

DSS is designed to be a meta-distribution framework, allowing creation of derivative distributions, both live or in standard package, built up upon a pure Debian release in a very simple way. In this way, it permits a high level of customization and easy maintenance, keeping the new application deployment/removal process fast and centralized. This feature is provided thanks to a special modular file system design, made possible by the adoption of UnionFS (see Wright et al., 2004 and Zadok et al., 2000).

UnionFS is a particular file system designed to merge different devices. DSS uses UnionFS to group physical devices with virtual (ramfs) ones to set up a final root filesystem. In virtual

A device in a GNU/Linux operating system is a reference to a physical or virtual medium, for example, a hard disk partition. Devices are used to perform all operations allowed by the Linux kernel. Almost all hardware supported by the Linux kernel has its specific device, including removable media like CDROM, USB disks and so on. A hard disk partition has its own device that must be used in order to access the file system contained in that partition. This operation is called "mounting." Usually, a mount operation associates a directory (a "mount point") to a device, and any further modification within the directory will be actually performed on the file system related to the device.

UnionFS allows mount operations involving an arbitrary number of devices into one mount point (see Illustration 2: UnionFS). UnionFS performs a virtual "merge" of file systems considered, as shown in Illustration 1: An example of file system merging with UnionFS. We refer the reader to the following sections for a more detailed description of features.

The DSS file system is split into modules (or layers), which are added together via UnionFS.All modules in DSS are compressed archives, which can be mounted at runtime as small file systems. These modules contains programs and libraries, which are merged together into a unique file system, thanks to UnionFS; additive module management permits the creation of a final file system layout which is essentially a "sum" of every used layer. A side effect of DSS modularization is the flexibility in final result of the merging process, which actually allows different installation profiles, based on a combination of a common set of modules.

Illustration 1. File system merging with UnionFS

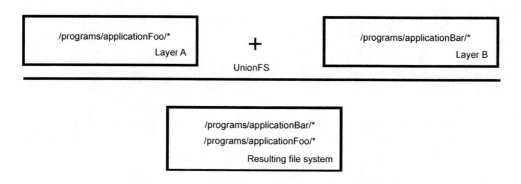

Illustration 2. UnionFS

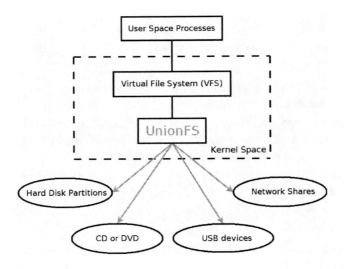

Illustration 3. The Upstream Salmon Struct (USS)

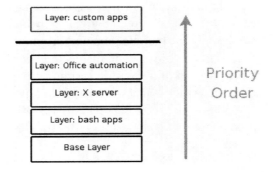

In Illustration 3: Upstream Salmon Stream (USS) structure for a desktop system, a possible combination of modules for a DSS distribution for a desktop PC is shown. Every module is created to provide a specific group of functionalities. A strict encapsulation of features allows a kind of "polymorphism" in resulting distribution. In this way, it is possible, for example, to generate a "Web server" distribution, starting from the set of layers shown in Illustration 3, including a layer with the Apache Web server and a DBMS, excluding the graphical server layer. In the same way, a customized GNU/Linux desktop distribution may be generated, containing a specific ready-to-use[5] corporative environment. An example of a possible usage of DSS in a corporate network may be seen in Illustration 4: DSS in a corporate

Illustration 4. A corporate network with DSS

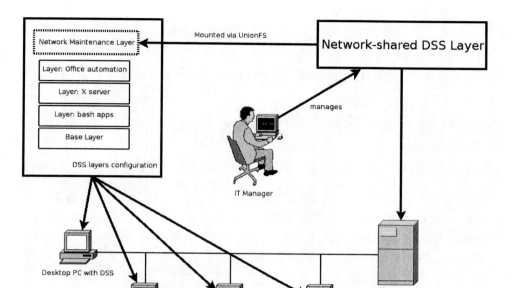

network. For a wider analysis, we remand to following section, dealing with "Upstream Salmon Struct."

Moreover, as compressed modules are merged in an ordered way, a single installation may be multi-purpose, including or excluding any of the boot loader parameters. This feature is very important to contain different installation profiles in a single location, and it is extremely useful in a network installation, or in a DVD release, for example.

Module creation process is also very simple, and it may be created in two ways: interactive and non-interactive processes.

Starting with the latter, it relies on a standard Debian tool (the "debconf" program, used for Debian distribution software management), just providing as input a list of desired Debian packages to be included in final module; a script would download, configure and re-pack packages into a compressed archive which is actually the desired module.

The interactive way, the simplest, needs only to boot DSS, and then, after using the usual GUI and front-ends for new software installation, invoke another script.

Resulting archives may be redistributed inside a standard DSS release without any further modifications to original status.

Technical Description of Features

Key Technology: "Early User-Space"

"Early user-space" mode is essentially a technicality, based on initramfs, a chunk of code that unpacks a compressed file system image (in CPIO format, see GNU Project, 2004), during the kernel boot process. It replaces the old initrd file system format, which contained a set of kernel modules stated to be available at boot time, before mounting the root file system and thus before having all kernel resources available. The main advantage of initramfs is its capability to be used with ramfs, a file system designed to work on physical RAM, scalable in size, instead of the usual initrd.

Early user-space allows DSS, in conjunction with later-described UnionFS, to save time in the boot phase: Instead of setting up a boot environment for hardware detection/configuration operations, DSS directly sets up a final working environment, and when the kernel finishes its startup operations, the boot process is over, with a simple environment update. This is because RAM allocated since boot start for required boot operations does not need to be freed/removed, and the previously adopted special environment (which is usually adopted for embedded systems) is not used anymore except for the boot process. Eventually, it is possible to allocate all available RAM on the system to improve overall performance, reducing physical medium access delays by copying the whole DSS content image into a memory partition.

Key Technology: UnionFS

UnionFS is a stackable file system that operates on multiple underlying file systems (see Illustration 2: UnionFS). It merges the updated contents of multiple directories but keeps their original physical content separated. The DSS implementation of UnionFS merges the DSS RAMdisk with the read-only file systems on the boot CD so it is possible to modify any read-only file as if it was writable. UnionFS is part of FiST, File System Translator project. Its goal is to address the problem of file system development, a critical area of operating-system engineering. The FiST lab notes that even small changes to existing file systems require deep understanding of kernel internals, making the barrier to entry for new developers high. Moreover, porting file system code from one operating system to another is almost as difficult as the first port.

FiST, developed by Erez Zadok and Jason Nieh in the computer science department at Columbia University, combines two methods to solve the above problems in an innovative way: a set of stackable file system templates for each operating system, and a high-level language that can describe stackable file systems in a cross-platform portable fashion. The key idea is that with FiST, a stackable file system would need to be described only once. Then the code-generation tool of FiST would compile one system description into loadable kernel modules for different operating systems (currently Solaris, Linux, and FreeBSD are supported).

UnionFS allows DSS to virtually merge (or unify) different directories (recursively) in a way that they appear to be one tree; this is done without physically merging the directories content. Such "namespace" unification has a benefit in allowing the files to remain physically separate, even if they appear as belonging in one unique location. The collection of merged directories is called a union, and each physical directory is called a branch. When creating the union, each branch is assigned a precedence and access permissions (i.e., read-only or read-writable). UnionFS is a namespace-unification file system that addresses all of the known complexities of maintaining Unix semantic without compromising versatility and the features it offers. It supports two file deletion modes that manage even partial failures. It allows efficient insertion and deletion of arbitrary read-only or read-writable directories into the union. UnionFS includes in-kernel handling of files with identical names, a careful design that minimizes data movement across branches, several modes for permission inheritance, and support for snapshots and "sandboxing."

UnionFS has an n-way, fan-out architecture (again, Zadok et al., 2000 and Wright et al., 2004). The benefit of this approach is that UnionFS has direct access to all underlying directories or branches, in any order.

DSS Inside UnionFS

DSS, just before performing the setup later described as "Upstream Salmon Struct," mounts different compressed file systems in different mount points and uses a read-writable directory as last layer, in order to prepare the environment to be hosted in just one final mount point (the root directory).

Even if the concept of virtual namespace unification appears simple, there are three key problems that arise when using it as root file system of DSS.

The first is that two or more unified directories can contain files with the same name. If such directories are unified, duplicate names must not be returned to user-space for obvious reasons. UnionFS solves this point by defining a priority ordering of the individual directories being unified. When several files have the same name, files from the directory with higher priority take precedence.

The second problem relates to file deletion. Files with same name could appear in the directories being merged or files to be deleted residing on a read-only branch. UnionFS handles this situation by inserting a without, a special high-priority entry that marks the file as deleted.

When file system code finds a without for a file, it simply behaves as the file does not exist.

The third problem is relegated to the previous one and it involves mixing read-only and read-write directories in the union. When users want to modify a file that resides in a read-only branch, UnionFS performs a "copy-up," that is, the file is copied to the higher priority directory and modified there.

UnionFS and the Upstream Salmon Struct (USS)

The power of DSS resides in its design, offering high modularity and allowing customization as easy as possible. This has been achieved by designing the USS and using UnionFS as background.

The unified root file system is comprised of the content of different modules, each a squashfs compressed file system (see Illustration 3: The Upstream Salmon Struct):

1.   **base:** Console mode module, it contains a basic bootstrapped Debian system.

2.   **kernel:** It contains the /lib/modules/ directory plus kernel related utilities.

3.   **xserver:** Graphical mode modules (in case file names clash, the priority in the unified directory is defined by sorting the modules name).

4.   **deliver:** It contains the runlevel scripts needed to reconfigure "debconf" database and the environment reading the user configuration from /proc/cmdline passed to kernel at boot from boot loader (e.g., locales information, force screen resolution).

5.   **overall:** The read-writable branch, it can reside in RAM or even be an external hd; base, kernel and xserver use is self-explanatory, but the packages inside those modules are stored using a "noninteractive" debconf frontend, and so they maintain their own default configurations; that is why DSS can be considered a pure Debian system booting from a CD/DVD/USB. To allow the user his or her own locales settings (i.e., language, keyboard) and video card optimized drivers, some packages need to be reconfigured, and this is done using the runlevel scripts in deliver.

Deliver

The scripts in "yuch-bottom," the directory within the initramfs, write the environment variables in the file /etc/deliver.conf, parsing command line parameters from boot loader, as lang(uage), username, hostname, and so forth. Deliver uses those variables to reconfigure some packages, upgrading at the same time the debconf database.

The scripts in deliver are plain text bash scripts, this allows DSS use not only for a i386 live distribution, but even for powerpc or sparc computers, and all the other 11 architectures that Debian supports, making DSS fully architecture-independent.

Thanks to its scripts, to be ported from an architecture to another DSS just needs the right initramfs and deliver module, without kernel customization, as it is sufficient as a pure Debian stock kernel.

DSS, as opposed to knoppix, uses debconf to configure the system, which provides a consistent interface for configuring packages, allowing the choice from several user interface frontends. It supports even a special "pre-configuration" of software packages before they are actually installed, which allows massive installation or upgrade sessions demanding all necessary configuration information up front, without user interactions (frontend "non-interactive"). It allows for the skipping over of less important questions and information while installing a package but providing a chance to revise them later.

It is also interesting to note that debconf itself is completely a Debian-supported tool, and its use is not customized at all: another key point in 100% Debian compatibility.

DSS and Usability

As said, DSS is mainly aimed at desktop systems and the improvement of global usability. This feature is very important, and we faced many problems when introducing DSS into our use case because of the experience of users with win32 systems. Our approach in that situation was twofold: We improved knowledge of users and we developed some ad hoc solutions.

Considering our interaction with testers, we planned to work on their formation, exploiting differences between Win32 systems and our distribution, to simplify their approach to DSS. As mentioned, due to the small number of users involved, our test may not provide general considerations. However, in this process we followed indications and recommendations coming from precedent experiences in larger contexts: in the region of Spain Extremadura (see Vaca, 2005 and ZDNet, 2005) for example, but also from case studies held in South Africa (see Brink et al., 2006). We also exploited our experience in training and documentation design, gained with several analyses of precedent free open source software adoption experiences (see, e.g., Rossi, 2006; Zuliani, 2004a, 2004b; Russo, 2005).

On the technical side, we worked to improve overall usability, exploiting early user-space features to manage the whole desktop system with just user-space software, as it is common to mount devices, for example.[6]

Indications from research activities advise addressing usability matters as crucial for easing the diffusion of free/open source software and how to properly implement strategies to simplify the interactions with users (see Nichols, 2001, 2004; Frishberg et al., 2002).

A remark on the principal difficulty in this process: As DSS is a meta-distribution, we could not rely on almost anything connected with window managers' capabilities. This means that what we could use was limited to the X graphical server and a few other software applications. One of them is the X notification daemon, commonly included in X distribution packages.

We added a set of scripts, rules, and policies to hardware management software, such as the largely adopted hardware abstraction layer software (HAL) and message deliver DBUS, to react to events like plugging in and out of USB devices, mounting and un-mounting them automatically in case of storage devices, or proposing a set of actions in the case of a newly inserted CD or DVD. Our notification is graphical in a standard notification window, and comes with a synthesized voice that reads the title of the notification, thanks to standard Debian-provided software (festival, Taylor et al., 1998). Thanks to HAL and DBUS features, it is possible to detect almost every hardware event, and this offers a lot of possibilities for extending default reaction behavior.

As for every other DSS component, this is script-based and completely customizable by the user or system administrator. Unluckily, this feature is not yet ready to come into production environments, and we are still bug-fixing and enhancing it.

The Community-to-Community Experience

During the design and implementation process of DSS, there were many contacts with other software communities, providing help to each other.

One of our greatest connections is with the team of developers of the Ubuntu community, and we acknowledge their great work and the easiness of the relationship. In fact, we had a continuous mail exchange, especially when they created their live installation program, ubiquity. They happily agreed to make some modifications to its behavior, and in this way we were able to include it in our mainstream, so now it also works for DSS. Another great point of interest is represented by interactions with people from File systems and Storage Lab at Stony Brook University, the creators of UnionFS and a lot of theories on file systems. They accepted our advices on UnionFS runtime features, as it was in heavy development phase when we adopted it, and both of us received an added value from this.

But we were also providers for other communities: Elive[7] developers adopted DSS as the starting distribution for their releases from 0.5 (see Distrowatch, 2006). In this way, as their goal is to care about Enlightenment window manager (Enlightenment community, 2006), they essentially created a layer to be added at top of the USS, without worrying about low-level details. This is one of the the first applications of meta-distribution features, one of our greatest aims in this project.

We acknowledge Daniele Favara, our distribution manager and main developer, for all these successful contacts and interactions. What is reported here is just a summary of all contacts Daniele kept, and we report them as the basisfor some considerations.

We may look at our experience as an example of positive integration of communities. DSS may be seen as a conjunction point between different groups with, obviously, different objectives, and where it may be very complicated to find an equilibrium point. In this case, we, and especially Daniele, were lucky and skilled in finding a good alchemy between pursuing "client" needs and "provider" objectives. The UnionFS team found an interesting application (a good "client") in DSS, while Elive developers chose DSS because it is a good "provider," which allows them to concentrate on their "core business," the window manager aspects.

Everyone gained from this relationship in terms of existing solution reuse, bug reporting, and, in some cases, code contribution. This may be viewed, in some aspects, as a demonstration of statements of Raymond in Cathedral and Bazaar (1999), as the famous "more eyeballs" that allow prompt bug reporting, as well as for role exchange between users and developers of an open source software project.

Not in every case it is possible to reach such a happy ending: It happened that several mail exchanges were started, proposing or receiving suggestions or modifications, but it was not possible to reach an equilibrium point. Sometimes there was a domain-specific requestwhich would lead to a loss of generalization to the "provider" community, or sometimes, unmatchable visions between proposers and developers.

Anyway, just relying on our experience, it is possible, once a "proper" set of providers and clients is found, to lead a parallel development process, practically constituting a whole, in which every partner receives more benefits than losses. This happens when it is possible to find the proper equilibrium point between features requested and provided, and with continuous attention to other communication channels, such as partner communities, developer mailing lists, forums, wikies, and so forth to keep each other in touch in a constructive way.

A Legal Aspect: Proprietary Software and DSS

DSS, as said, is a completely 100% Debian-based distribution, and for this reason it strictly follows Debian Guidelines (see Debian, 2004). So, it is not legally possible to distribute proprietary-licensed software in conjunction with DSS.

But we acknowledge that it is common, in public administration and in some companies, to deal with proprietary software.

Our choice is, again, to follow Debian guidelines, and so forth allowing and supporting the creation of layers containing non-free software, but permitting their delivery/deployment in separate archives and repositories, following the "non-free" software behavior of Debian distributions. Even if DSS requires a separate delivery mechanism, it completely supports non-free software, without any further impact or differentiation.

This is a key point. In this way it is possible to meet users requirements and license restrictions at the same time.

Future of DSS

DSS is in continuous development, and it is very difficult to give a summary of development directions. One of our greatest points of interest is to make the production of custom layers even easier and more automatic than today. We are looking at specific wiki syntax to give directives to a script able to automate the creation process of a DSS layer; in this way, the editing a wiki page alone would result in an automatic generation of a layer ready to download and use, without any other user activity.

We are also focused on usability features: Our notification system still needs testing and tuning, for example. Thus, we plan to involve some usability experts to improve it and to suggest further objectives.

We also trust that from a forthcoming large-scale migration test, in which we will try to apply lessons and recommendations coming from the growing academic literature, we may be able to obtain useful directions for PAs and companies willing to embrace an analog path.

Conclusion

DSS is a 100% Debian live distribution, and may be proficiently used to install a pure Debian system on a desktop PC. Thanks to its features, it is very simple to customize as a starting base version, in a way to meet, for example, large-scale installations with specific requirements, such as in large company networks. Its maintenance is not effort-prone, due to the adoption of standardized technologies. Thanks to the DSS innovative design, the software represents a unicum in the current scenario. Moreover, there are no limitations to port DSS into any of Debian supported architectures, or to use it in embedded systems. We acknowledge that one of our greatest successes may be found in relations with other communities, which allowed us to improve DSS and to help our partners at the same time.

Acknowledgment

We acknowledge Daniele Favara, who is the main developer and the community leader of DSS. Without his great job DSS would not exist.

References

Assinform. (2006). *Report on IT, TLC and multimedia 2005.* Assinform.

Brink, D., Roos, L., Weller, J., & van Belle, J.-P. (2006). *Critical success factors for migrating to OSS-on-the-desktop: Common themes across three South African case studies.* Boston: Springer.

COSPA Project. (2003-2006). *Work package 4, deliverable 4.3—Experience report on the implementation of OS applications in the partner PAs.* Retrieved August 14, 2006, from http://www.cospa-project.org/download_access.php?file=D4.3-ExperienceReportOnTheImplementationOfOS.pdf

Crowston, K., Annabi, H., & Howison, J. (2003). Defining open source software project success. In *Proc. of Int. Conference on Information Systems.*

Debian. (2004). *The Debian free software guidelines.* Retrieved January 2005, from http://www.debian.org/social_contract#guidelines

Distrowatch. (2006). *Elive 0.5 Beta 3.1 Screenshot Tour.* Retrieved August 15, 2006, from http://osdir.com/Article9182.phtml

Enlightenment community. (2006). *Enlightenment project.* Retrieved April 12, 2006, from http://enlightenment.sourceforge.net/

Frishberg, N., Dirks, A. N., Benson, C., Nickell, S., & Smith, S. (2002). *Getting to know you: Open source development meets usability extended abstracts of the conference on human factors in computer systems (CHI 2002)* (pp. 932-933). New York: ACM Press.

Free Software Foundation. (1991). *GNU GENERAL PUBLIC LICENSE v2.* Retrieved April 12, 2006, from http://www.gnu.org/licenses/gpl.html

GNU project. (2004). *The CPIO project.* Retrieved April 12, 2006, from http://www.gnu.org/software/cpio/

Hotplug community. (2001). *Hotplug project documentation.* Retrieved April 12, 2006, from http://linux-hotplug.sourceforge.net/

KDE e.V. (2006). *KDE Documentation.* Retrieved April 12, 2006, from http://www.kde.org/documentation/

Knopper K. (2000, October 10-14). Building a self-contained auto-configuring Linux system on an iso9660 file. In *Proc. of the 4th Annual Linux Showcase and Conference*, Atlanta, Georgia (pp. 373-376).

Murdock I., (1994). Overview of the Debian GNU/Linux system. *Linux Journal*, (Vol. 1994, 6es).

Nichols, D. M., Thomson, K., & Yeates, S. A. (2001). Usability and open source software development. In E. Kemp, C. Phillips, Kinshuck, & J. Haynes (Eds.), *Proceedings of the Symposium on Computer Human Interaction* (pp. 49-54). Palmerston North, New Zealand: SIGCHI New Zealand.

Nichols, D. M., & Twidale, M. B. (2003). The usability of open source software. *First Monday*, 8(1).

Petullo, M. (2005). Encrypt your root filesystem. *Linux Journal*, 2005(129).

Rossi, B., Scotto, M., Sillitti, A., & Succi, G. (2006). An empirical study on the migration to OpenOffice.org in a public administration. *International Journal of Information Technology and Web Engineering*, 1(3), 64-80.

Russo B., Sillitti A., Zuliani P, Succi G., & Gasperi P. (2005, May 15-18). A pilot project in PAs to transit to an open source solution. In *Proc. of The National Conference on Digital Government Research (DG.O2005)*, Atlanta, Georgia.

Rustling, D. (1999). *The Linux kernel.* Retrieved April 12, 2006, from http://www.tldp.org/LDP/tlk/tlk.html

Stallman, R. M. (2002). *Free software, free society: Selected essays of Richard M. Stallman.* GNU Press.

Suzak, K., Iijima, K., Yagi, T., Tan, H., & Goto, K. (2005). SFS-KNOPPIX. In *Fourth IEEE International Symposium on Network Computing and Applications* (pp. 247-250).

Taylor, P., Black, A., & Caley, R. (1998). The architecture of the festival speech synthesis system. In *Third International Workshop on Speech Synthesis*, Sydney, Australia (pp. 147-151).

Ubuntu group. (2004). *Ubuntu philosophy.* Retrieved April 12, 2006, from http://www.ubuntu.com/ubuntu/philosophy

Vaca, A. (2005). Extremadura and the revolution of free software. In M. Wynants & J. Cornelis (Eds.), *How open is the future? Economic, social & cultural scenarios inspired by free and open source software* (pp. 167-197). Brussels, Belgium: VUB Brussels University Press.

Wright, C. P., Dave, J., Gupta P., Krishnan H., & Zadok E. (2004). *Versatility and Unix semantics in a fan-out unification file system.* Retrieved June 15, 2006, from http://www.fsl.cs.sunysb.edu/docs/unionfs-tr/

Zadok E., & Nieh, J. (2000). FiST: A language for stackable file systems. In *Proc. of 2000 Usenix Annual Technical Conference* (pp. 55-70).

ZDNet. (2005). *Extremadura Linux Migration case study.* Retrieved June 15, 2006, from http://insight.zdnet.co.uk/software/linuxunix/0,39

Zuliani, P., & Succi, G. (2004a). An experience of transition to open source software in local authorities. In *Proceedings of e-challenges on software engineering.*

Zuliani, P., & Succi, G. (2004b). Migrating public administrations to open source software. In P. Isaías, P. Kommers & M. McPherson (Eds.), *Proceedings of e-society 2004 IADIS international conference* (pp. 829-832). IADIS Press.

Endnotes

[1] http://www.dsslive.org

[2] http://www.knopper.de

[3] This is acknowledged by Klaus Knopper himself, the creator of Knoppix, in the Knoppix support forum, excluding the possibility to update a Knoppix box with standard Debian procedure ("apt-get dist-update")

[4] See Eckert, J. W., & Schitka, M. J. (2004). *MCSE guide to managing a Microsoft Windows Server 2003 Network.* Thomson Learning

[5] There may be implications, depending on license of software delivered within layers: the Debian Guidelines (Debian, 2004) explicitly prevents redistribu-

tion of software with a license not compliant with free software ones; see next paragraphs

[6]   In traditional GNU/Linux distribution, operations dealing with devices may be performed only by super-users. User-space tools are effective to allow some of those operations to users, in a safe way.

[7]   Elive distribution, from its website http://www.elivecd.org/, is a Debian-based distribution, powered by Enlightenment window manager (http://enlightenment. sourceforge.net/) and with a strong accent on usability aspects, but focalized primarly on window manager tuning

Section V

F/OSS Case Studies

Chapter XI

Assessing the Health of an Open Source Ecosystem

Donald Wynn, Jr., University of Georgia, USA

Abstract

This study examines the concept of an ecosystem as originated in the field of ecology and applied to open source software projects. Additionally, a framework for assessing the three dimensions of ecosystem health is defined and explained using examples from a specific open source ecosystem. The conceptual framework is explained in the context of a case study for a sponsored open source ecosystem. The framework and case study highlight a number of characteristics and aspects of these ecosystems which can be evaluated by existing and potential members to gauge the health and sustainability of open source projects and the products and services they produce.

Introduction

Open source software (OSS) products are an increasingly significant part of the IT infrastructure for many enterprise customers. The combination of lower licensing fees and increased technological flexibility is a seductive lure for corporate administrators looking for effective solutions for their operations. However, enterprise customers are often not willing to entrust their mission-critical applications to software and systems that are unproven and unsupported. These customers typically require high levels of customer assurance and on-demand support in order to guarantee the continuous, perpetual operation of their firm's information technology architectures. The value of information technology investments is sustainable to the extent that the vendor can continue to attract and support a network of skilled professionals and organizations that are motivated to provide the necessary functions including product development, distribution, service provisioning, and marketing. Customers who deploy technology products and/or platforms that fail to attract a thriving ecosystem or whose ecosystem deteriorates (e.g., IBM's mainframe customers) are increasingly faced with declining availability of skills, increasing operating costs, and/or lower levels of innovation (Vecchio, 2004).

The prototypical OSS product is described as one in which the developers are primarily volunteers. The software emerges from a loosely coordinated, unsupervised community of developers and other contributors who band together for reasons including self-interest, profit motives, and protest against large closed-source software providers (Hars & Ou, 2002; Hertel, Niedner, & Herrmann, 2003). These communities typically have little or no formal organization or corporate structure. This makes it difficult for most enterprise software customers to deploy open source solutions, since there is little or no means of ensuring the resources that are necessary for long-term support and maintenance.

As a result, the structures and relationships surrounding many open source software projects have evolved in order to address these needs. "Professional Open Source" companies such as JBoss (Watson, Wynn, & Boudreau, 2005) and MySQL offer a wide range of support options, including 24x7 phone support, while continuing to distribute their products under an open source license that typically allows customers to use the software without the upfront licensing costs.[1] In return, these firms are able to ensure higher levels of dedicated support and development by hiring their developers and support staff and by contracting services with third parties. Alternatively, sponsoring organizations are able to dedicate specific resources (including employees and financial assets) to the project to more tightly control the development and support of software products. In both cases, the open source license both supports and encourages participation by other individuals and organizations in the innovation and support functions. It is from the participation of this complex web of partners, distributors, consultants, and other third-party entities, as well as the original developers, that users are able to augment the functions provided by the software vendor.

We characterize these individuals and organizations as members of a larger *ecosystem* that includes the original vendor as well as supporting foundations (O'Mahony, 2005), external service partners, integrators, distributors, and the users themselves (Krishnamurthy, 2005). As such, this ecosystem encompasses the full range of effort done on behalf of, in support of, and as a result of the software being created. The term *ecosystem* is a frequent part of the vocabularies of both the popular press (Lacy, 2005) and software vendor executives with firms such as IBM, SAP, JBoss, MySQL, and others. While increasing in popularity, the term is often applied in different contexts depending on the speaker. To date, there has been little focused research on open source software ecosystems despite pronouncements that the resultant software products reflect the health, maturity, and stability of the community that surrounds them (Woods & Gauliani, 2005). There has also been little research focused on the components and qualities of these systems that enable them to sustain themselves.

In this chapter, we will develop the ecosystem concept beginning with a brief examination of its origins from ecology. We apply these foundations toward a definition of an open source software ecosystem and the derivative concept of ecosystem health as a proxy for the sustainability of the social system which surrounds the OSS project. We then use these definitions to identify the determinants of the health of the ecosystem. These determinants include the individual and organizational units comprising the ecosystem, the relationships between them, and the practices enacted by the various units individually and collectively. In order to ground our study, a case study methodology is used to examine the health of the ecosystem surrounding an emerging publicly sponsored open source project, along with vignettes from other forms of open source projects.

Ecosystem Concept

Ecosystem theorizing in ecology began as researchers realized the importance of the relationship between a community of living units and the abiotic, or nonliving, units and environment in which they existed (Tansley, 1935). As the concept matured, it included theorizing on the flow of energy as well as materials (Lindeman, 1942). An ecosystem can be defined in a number of ways (see Table 1). However, at its core, an ecosystem "involves the circulation, transformation, and accumulation of energy and matter through the medium of living things and their activities" (Evans, 1956). This definition includes the flow of materials (such as chemical elements, water, pesticides, etc.) and energy (sunlight, heat, etc.) among the various plants,

Table 1. Ecosystem definitions

Definition	Citation
Ecology & Human Ecology	
"The whole system (in the sense of physics) including not only the organism-complex, but also the whole complex of physical factors forming what we call the environment of the biome-the habitat factors in the widest sense." (299)	Tansley (1935)
"An ecosystem is an arrangement of mutual dependencies in a population by which the whole operates as a unit and thereby maintains a viable enviromental relationship." (26)	Hawley (1986)
Business Ecosystems	
"An economic community supported by a foundation of interacting organizations and individuals—the organisms of the business world." (26)	Moore (1997)
"The relationships among a community's populations, between the community and its environment, and, by extension, among their resource supplies and among their resource suppliers form the community's ecosystem." (87)	Bidwell and Kasarda (1998)
Like business networks, biological ecosystems are characterized by a large number of loosely interconnected participants who depend on each other for their mutual effectiveness and survival. (8)	Iansiti and Levien (2004)

animals, and microorganisms in an ecosystem as well as non living components such as water, air, soil (Chapin, Matson, & Mooney, 2002). Despite these central tendencies, the concept of an ecosystem can be difficult to define clearly, even among scholars in ecology—its native discipline. This ambiguity has not discouraged the use the ecosystem concept, which has proven to be useful in several dimensions (Pickett & Cadenasso, 2002). As a metaphor, ecosystems provide a rich, informal term which allows stakeholders to easily view their roles in a holistic view of a given network of relationships.

Parallel to the development of the ecosystem concept as a tool for understanding ecological systems, a similar concept was being developed in the field of human ecology. An offshoot of sociology, human ecology was introduced as a means for investigating the symbiotic relationships within a community of humans (McKenzie, 1934). The ecosystem concept was ported over from ecology, particularly from the study of plant communities (Hawley, 1986). From this analogous beginning, human ecology employed many of the spatial and organizational concerns of bio-ecological systems to investigate ways that human populations attempted to sustain their existence in a specific environment.

Human ecology defines an ecosystem by the units, relations, and functions that comprise it (Hawley, 1986). Units are the entities that exist in the environment under investigation. These can exist in either simple units (individuals) or complex units (organizations or groups). Relations refer to the interdependencies between the units, while functions are those activities that produce materials, distribute them among the various units, and coordinate the other functions to ensure sustenance and efficiency. As defined, this resembles the "circulation, transformation, and accumulation" functions identified for biological ecosystems above. However, human ecologists often include the transfer of information along with materials and energy.

The interdependent relationships that exist between units (including both individuals and organizations) in an economic community bear a strong similarity to the preceding discussion of ecosystems in ecology and human ecology. The units in a business ecosystem interact with each other in order to perform a set of functions which enable the materials and information to be acquired from the environment and then circulated, transformed, and accumulated by other units. These functions are based around an existing set of goals which are the reason that the ecosystem exists.

As expected, representative definitions for business ecosystems (see Appendix A) include many of the same conceptual elements found in ecology and human ecology. However, instead of referring to materials, energy, and information as in these ecosystems, we assume the operation of the units involved in an economic community in terms of various forms of capital. Thus, a business ecosystem can be defined as *an arrangement of individual and organizational units involved in or influencing the circulation, transformation, and accumulation of capital (in various forms) through their functional activities.* In order to fit the specifics of the particular context of our study, we modify the definition of open source software ecosystems as follows: *an arrangement of individual and organizational units, involved in or affecting the circulation, transformation, and accumulation of capital (in various forms) in order to provide cooperative development, testing, marketing, distribution, implementation, and support of open source software.*

Ecosystem Health

As stated above, a key characteristic of these ecosystems with respect to their suitability for long-term adoption for critical IT infrastructures is their ability to sustain themselves. In essence, an ecosystem that is capable of producing and circulating sufficient levels of capital to induce participants to contribute resources and effort can be considered to be a *healthy* one. Unhealthy ecosystems are those that fail to satisfy the necessary returns to each member. Specifically, ecosystem health refers to the ability of an ecosystem to remain viable and capable of sustaining the participation of the members that comprise it. Analogously from ecology, the concept of ecosystem health is multidimensional, encompassing the vigor, organization, and resilience of an ecosystem (Costanza & Mageau, 1999; Mageau, Costanza, & Ulanowicz, 1995). More specifically, this construct implies that the system can maintain its structure (organization) and aggregate function (vigor) over time in the face of external stress (resilience). A similar set of concepts have been applied to business ecosystems as niche creation, productivity, and robustness (Iansiti & Levien, 2004), which are similar in definition to organization, vigor, and resilience, respectively. We now look at these dimensions in more detail, with brief examples drawn from the PostgreSQL project.[2] PostgreSQL is an open source relational database that has been in existence over 15 years and has proven to be a stable, functional development and support community.

Organization

The organization of an ecosystem includes the diversity of units and the interactions between them. A highly organized ecosystem will consist of a highly diverse set of specialists with an equally diverse set of exchange pathways (Costanza & Mageau, 1999). A less organized ecosystem will have fewer unit types and/or fewer interactions between them. Note that the unit diversity is not necessarily a direct substitute for the number of interactions between them. The key lies in the capacity of an ecosystem to efficiently communicate and transport resources as the units diversify into specialized providers.

The diversity of business ecosystems includes core contributors, suppliers, partners, complementary vendors, distributors, customers, and even competitors. The interaction between these units is determined by the existence and strength of formal and informal arrangements, norms, intraorganizational structures, and coordination mechanisms. Measuring these interactions is a complex task, especially for open source ecosystems in which they are often informal, highly unstructured, and temporally unstable. Organized ecosystems are characterized by more structured, formal, and stable relationships.

The PostgreSQL community consists of a core development team that is responsible for both the general strategic direction and day-to-day operations and development of the project, an additional layer of developers that coordinates the development of other modules and patches, and a number of other contributors who may submit patches and bug fixes for other developers to evaluate. Additionally, there are a number of additional contributors including sponsors, support partners, and end users that contribute capital and resources toward the maintenance and growth of the PostgreSQL community. There are several communications paths available to the members, including the primary Web site, several mailing lists, an IRC channel, a newsletter, and a bugtracking application. As a community-based project, PostgreSQL uses many of the norms typically associated with open source, including the open access to code, shared bug-fixing, and active user-to-user support and maintenance forums. The specific organizational structures of this community both support and sustain these norms as well as enable the production of both software and services.

Vigor

The purpose of an organization is to combine the contributions of a number of individuals and groups of individuals toward a specific set of aims. These contributions, as well as the resulting inducements, can be classified as different forms of capital, including *social* (Adler, 2001; Bourdieu, 1985; Coleman, 1988; Lin, 2001; Nahapiet & Ghoshal, 1998), *symbolic* (Bourdieu, 1985), *human* (Becker, 1993), *organizational* (Youndt, Subramaniam, & Snell, 2004), and *economic* capital. Human capital is defined as the skills, knowledge, and abilities of an individual that can be used to generate income or other useful outputs (Becker, 1993). There are three dimensions of social capital: structural, cognitive, and relational (Nahapiet & Ghoshal, 1998). The accumulation of goodwill and prestige within a social system can be defined as a form of symbolic capital (Bourdieu, 1985). Organizational capital is the accumulation of routines and processes by an organization or group. The various units of capital are acquired, converted, combined, and distributed throughout the ecosystem by the units themselves. Vigor is essentially the aggregate level of this capital flow.

The vigor is the throughput or productivity of the ecosystem, which is analogous to the gross domestic product (GDP) as calculated in economics. This consists of the full range of outputs offered by the members of the ecosystem, and the capital resources contributed to produce them. The functional outputs produced in software ecosystems include the software and source code as designed and coded by the developers. It also includes services required by the other members of the ecosystem such as training, installation, testing, maintenance, and support. Additionally, the promotion and distribution of the software is included in the calculation of the vigor of the system.

As identified earlier, there are a number of units that are actively involved in the PostgreSQL community. A number of sponsors, including Sun Microsystems and Red Hat, provide financial capital and support to the development community as well as server space and other resources. Several firms provide support for the developers that provide code and other services for the community, although certainly many of the active contributors are not actively supported by any other unit. The capital resources contributed by these members are combined and aggregated to produce the software itself as well as the information and support that are the instruments through which the community sustains itself.

Resilience

The resilience of an ecosystem is its "ability to maintain vigor and organization in the presence of stress" (Costanza & Mageau, 1999). In other words, a resilient system maintains its exchange relationships in order to continue producing outputs following perturbations which the entire system or individual units may have encountered. The ecosystem may not return to its exact structure or pattern of organization after the stress, but may have to establish a new pattern to adapt to the changing environment.

Resilience includes both the ability to resist the effects of stress as well as the ability to recover from any ill effects that may result. In much the same way, civilizations such as Easter Island and the Roman Empire were able to resist certain stresses and recover from yet others, but ultimately collapsed in the presence of others (Diamond, 2005).

The specific stresses which an ecosystem can resist and recover may differ from one ecosystem to another. Two dimensions exist for stresses in ecosystems: impact and source. The impact of a given stress incident may be positive ("eustresses") or negative ("distresses") with respect to the effect of the stress on the vigor and organization of the ecosystem (Selye, 1974). The second dimension identifies the ultimate cause of the stress. Some events are generated by significant changes in the environment in which the ecosystem operates while others are the result of changes in the internal dynamics of the ecosystem itself.

PostgreSQL has experienced a number of positive and negative stresses over its history. For instance, in 2000 a dot-com era startup firm (Great Bridge) hired a number of the core team members and active contributors, effectively intending to commercialize the project and provide promotional efforts and a stable base of financial support for the entire project. However, Great Bridge (like many other dot-com firms of that era) failed to attract sufficient revenues and went out of business in 2001, which threatened to shut the complete PostgreSQL community down. However, the community was able to rebuild itself in time, eventually returning to a stable, thriving state.[3]

Conceptual Framework

This chapter employs a conceptual framework for ecosystem health as shown in Figure 1. The organization of the ecosystem consists of the structure and interactions between units of the ecosystem. Vigor is represented by the conversion of capital into the various output functions, which generate other capital benefits. These benefits are either appropriated by units external to the system or returned to the contribut-

Figure 1. Conceptual framework of ecosystem health

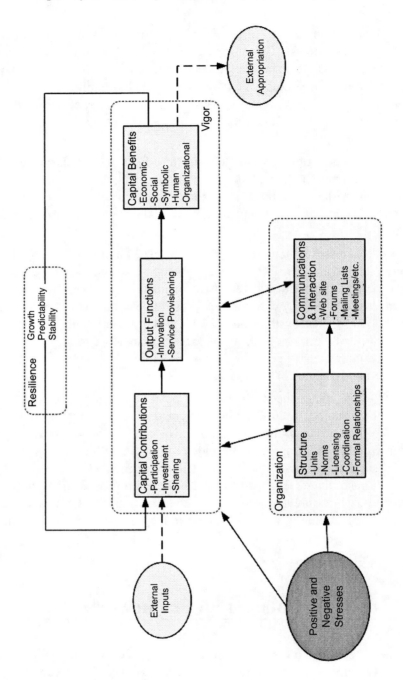

ing units for later use. Resilience exists to the extent that the ecosystem resists or responds to positive and negative stresses that affect its vigor and organization. The objective of this chapter is to follow this framework to explain the health of open source software ecosystems.

Research Design

An exploratory case study methodology was utilized to investigate the above conceptual framework in the study of an emerging open source software ecosystem. One of the strengths of case study research is its applicability in situations where the phenomenon of interest is being investigated "within its real-life context, especially when the boundaries between phenomenon and context are not clearly evident" (Yin, 2002, p. 13). Clearly our depiction of the flows of capital in an open source software ecosystem is deeply embedded in the given context. Therefore, the case study method allows us to more fully examine these activities as we experience them. The data for the case study were collected via semi-structured interviews with participants within the ecosystem, observations of meetings among the various participants, and access to existing archival data such as media articles, blog entries, mailing list archives, and discussion boards. The interviews typically lasted for an hour and were recorded and transcribed. During the observations, the researcher was not an active participant but a totally passive observer. The plentiful archival data about the two projects enabled us to reach a great deal of information about the operations and strategic decisions within the ecosystem, particularly from their most senior members. Once collected, the data were read and re-read to enable the researchers to become immersed in the data prior to analysis.

The case highlights an emergent form of open source ecosystems in which an organization, in this case a state government agency, sponsors the development of an open source software application primarily for its own benefit. In doing so, the agency is able to avoid vendor lock-in, which it currently is experiencing, and to ensure that its needs are specifically met. The case is examined in greater depth in the sections to follow.

Case Background

Evergreen is an open source project sponsored by the Georgia Public Library Service (GPLS). Currently, the library operates the PINES (Public Information Network for Electronic Services) system, which is a proprietary integrated library system

Figure 2. Evergreen ecosystem

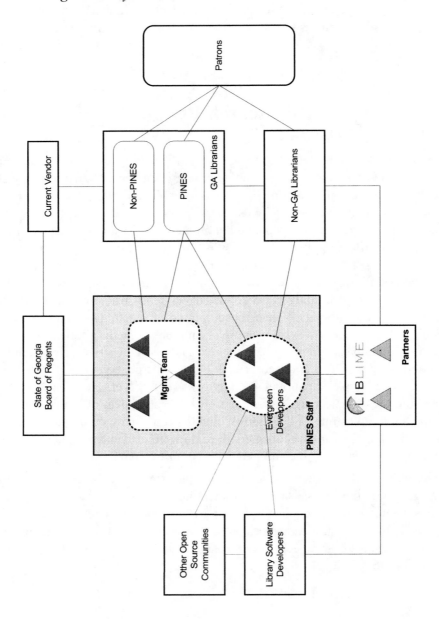

(ILS) that automates the circulation and cataloging functions for 252 libraries in 123 counties throughout the state. Under the current system, patrons can request materials from any of the consortium libraries without additional cost. However, the proprietary software is incapable of providing several of the functions needed by the library consortium, specifically those functions directly related to the local acquisition and system-wide circulation of new materials. The goal of the Evergreen is to replace the current ILS with one developed in-house (Open-ILS) under an open source license. In so doing, Open-ILS will not only address the shortcomings of the proprietary system, but also it will enable the state to save several hundred thousand dollars each year in licensing fees to the proprietary vendor and in support contracts required for the hardware requirements of the current software.

Not every library in the state is part of the PINES consortium. Several of the larger libraries, such as the Atlanta-Fulton County library system, are not members because of a concern that patron requests from the smaller libraries would consume many of the materials, reducing access by local patrons. These larger systems have the resources to select and operate separate software vendor contracts. Upon completion of the Evergreen project, several of these larger libraries may join the PINES consortium in order to share in the significantly reduced costs.

Organization

The core of the project includes the developers and the management team, all of whom are hired by the GPLS. Shortly after getting approval to launch, one developer was reassigned from another role on the PINES staff and two new ones were hired. Additionally, there are four staff persons and two interns working primarily on the project, including not only development but training, support, and project management. A contract was struck with developers at Liblime, an open source library software consulting form, to provide software testing and quality assurance services.

The three GPLS developers are responsible currently for nearly all of the development of Open-ILS. Some development is provided by the contract firm hired to perform QA tasks for the project, but there is no other major development work being performed by the users themselves (as is typical in open source projects). However, the current PINES librarians play an important part in ensuring that the developers are aware of all the tasks that are necessary in their daily routines. To provide this knowledge to the developers, the librarians have participated in several face-to-face meetings in their home locations and in the developers' offices. The local visits and focus groups were instrumental both for providing insight into the procedures and daily routines that the librarians follow and the new features and modifications that the librarians wanted. The periodic committee meetings are primarily used to update the librarians on the developers' progress and to ensure that the work done to date is satisfying their needs and requirements.

The most common means of communication between the librarians, PINES staff, and the development team is via e-mail. There are a number of restricted mailing lists that have been set up to enable communications to occur within specific subsets of the community, including the core team (PINES staff and development team), specific subcommittees (including circulation and cataloging), and all PINES members (core team plus subcommittees). In addition, two public mailing lists exist to facilitate technical and non-technical discussions, respectively. The public lists are frequented by the development team, partners, and interested observers, including librarians from outside the state of Georgia. Many of these external members of the ecosystem also communicate via a blog containing items such as status reports, corrections, and other announcements.

The development team members are all located in the local area, which enables them to meet once per week to discuss things such as code progress, staff matters, and plans for upcoming releases. Otherwise, they communicate primarily via an IRC channel that they inhabit for much of the day. They also monitor and contribute to an additional IRC channel used by the larger community of library software programmers.

Because of the perceived lower cost of open source compared to the vendor offerings, there is some degree of interest in open source library packages such as Evergreen and Koha. Users outside the state have been instrumental in providing technical assistance, bug reports, and suggestions based on their own needs and experiences, documenting many installation/support tips on the project Web site, and promoting awareness of the project among the library developers around the world.

Vigor

The contributions made to the ecosystem, as well as the benefits received from the ecosystem, can be expressed in terms of different forms of capital, which are convertible into other forms on the basis of various exchange relationships (Bourdieu, 1985). As stated above, the total production and circulation of capital within an ecosystem can be identified as the *vigor* of the ecosystem. In this section, we examine the vigor of the Evergreen system. The typology of capital employed here includes economic, social, human, symbolic, and organizational capital.

In many respects, the Evergreen project is designed to recapture economic capital that would otherwise be expended annually. Currently, the PINES system costs approximately $15M per year in licensing and maintenance fees. When complete, the Evergreen project will save the State over $10M per year. The project is funded directly through the GPLS, a unit of the University System of Georgia. These funds are used for hiring, facilities, equipment, and other expenses necessary for the operation. Because this financial support is responsible for the vast majority of the contributions made to the project, there is little incentive to tailor the code to any group outside of the PINES system. There are no plans to actively market the application to any other library system. Instead, the hope is that other library systems that wish to deploy the software will hire their own development and support teams to customize the software for their own needs (and contribute the changes back to the original project) while the current team of developers continues to support the PINES community.

As stated above, the vast majority of the application development is provided by the three GPLS-employed developers. However, other members of the ecosystem contribute human capital (knowledge, experiences, and values) in different ways. The staff and librarians share their experiences and knowledge of the current practices and workflows to the developers, who incorporate them into the new software. These librarians also provide information to each other in an effort to share useful hints and tips with respect to the use of the system. Outside users have taken the lead to write much of the documentation, which resides on a wiki site.

As stated above, there are three dimensions of social capital: structural, cognitive, and relational (Nahapiet & Ghoshal, 1998). Within the Evergreen ecosystem, structural contributions include access to external knowledge and resources via the multiple communications paths and the underlying hierarchical structures of the GPLS and PINES networks.

Evidence of contributions from the cognitive dimension includes the existence of shared language and codes. The developers had to obtain this shared knowledge by spending time with the librarians and staff members. However, they have contributed to this shared knowledge through new constructions in the new ILS as "meat grinders," which reflect a batch processing for "buckets" of books such as book bags or staff recommendation lists. Over time, the developers and users have forged a common language based on each other's individual terminologies.

The relational dimension includes factors such as trust, norms, and obligations. It is apparent that these factors have developed along with the application. Trust is the belief that the other party will act in an appropriate manner and is based on belief in the good intentions, competence, and reliability of the other party (Nahapiet & Ghoshal, 1998). The project manager explained the perceived trust PINES librarians have in the development team as follows:

It's almost like PINES has faith in us. They know we're gonna do the right thing. They know that we understand their needs. They know that [the PINES Project Director] has been in Georgia libraries for 20 years. They know that we know. We feel their pain. So, they feel like they already have a dog in the fight. They know that we're gonna do the right thing.

The norms of cooperation are evident in the amount of shared production that exists between the different roles, particularly between developers on the IRC channels.

Code is shared, testing is provided, and information is passed between the channel inhabitants with very little contractual obligation to comply. Instead, it is the expectation of future benefits from the other participants that appears to drive this sharing.

Along with the trust and obligations, the various members of the team have developed levels of symbolic capital in the form of legitimacy and social status as the project takes shape. As the software has taken shape, the heightened interest by outsiders has led to invitations for the staff and developers to make presentations about the project at library conferences and professional meetings throughout North America. This symbolic capital can be exchanged for other forms of capital at a later point (Bourdieu, 1985). For instance, the status an external individual attains because of his/her recognition as a member of the ecosystem may in fact be convertible into economic capital if GPLS needs to hire a new staff member.

Organizational capital is generated through the codified documents and procedures that arise over time as particular ways of working are routinized. The current processes reflect the ideal operation of the librarians as well as the capabilities of the underlying ILS software. Because the ILS is an integral part of many of the functions of the libraries' operations from online catalogs to patron fines to circulation reserves, the system heavily influences the actions of the librarians. It is these actions and procedures that the new system must account for and incorporate. These procedures are not always incorporated verbatim, especially workarounds that were created because of the limitations of the current system. The staff has been trying to convince the librarians to "try to think outside of the way we do things now." As such, many new policies may need to be changed as the new system is incorporated. Many of these policy changes must be approved by the PINES executive committee before being incorporated into the software. Thus, the organizational capital must be generated from the capabilities of the system but must be regenerated in many cases once the system is complete.

Functional Outputs

There are two primary functional outputs within the ecosystem: the Evergreen software and the services required by the other members of the ecosystem in support of the software. These services include training, implementation, troubleshooting, and more. We discuss these further in this section.

The Evergreen software is currently released in pre-beta, with the full rollout implementation to each of the libraries (as well as several others that are awaiting completion of the software) scheduled for later this year. The software is built on top of a number of other open source components. According to the developers, "part of the original rationale for going open source [as opposed to a proprietary solution] was the availability of OSS components such as Spidermonkey, XUL, and more." By utilizing the efforts of others, the developers were able to produce working code quicker and more efficiently. In return, the developers contribute code out to the greater community. One of the developers rewrote portions of the Jabber server while another wrote an interface using OpenSearch for the library OPAC. As such, the staff has established themselves as trusted members of the overarching library technical community.

The training will be provided originally by the PINES staff to the library directors. These directors (or their staff) will in turn provide the training to the rest of their staff. Because the system is designed to be intuitive and the various staffs are already used to a similar system, it is believed that the training will be much easier than the original training by the vendor of the current system. Similarly, the implementation processes will be provided by the librarians or their staff. The current system is supported by a help desk that will transition over to the new system. However, much of the help will be provided by other users via the mailing lists or via personal contacts.

Resilience

The final component of ecosystem health is resilience, which is the capacity to resist stresses that could affect the vigor and organization of the system and the ability to respond to changes that exceed this resistance (Costanza & Mageau, 1999). This resilience can be manifest as the sustainability or stability of the system, predictability in terms of the expected future outcomes, or growth in the capital circulated and accumulated within the system. For organizations, resilience depends upon several characteristics, including surplus resources, flexibility, commitment, coordination, and leadership (Horne III, 1997). Processes that promote resilience are typically emergent and adaptive (Sutcliffe & Vogus, 2003).

In the Evergreen project, resilience is enacted through the availability of resources to meet the challenges that arise. In many respects, the existence of the project is the end result of an emergent process that enabled GPLS to code the software in response to the perceived inability of the current vendor to meet its operational needs. Rather than continuing to operate in a manner that required constant workarounds and adjustments, the decision was made to create a system which could be tailored to fit the specific needs of a large library consortium such as PINES.

One of the possible stresses that may arise in this ecosystem is the loss of a developer.[4] There is a significant investment in training each of the developers to enable them to understand the processes embodied in the PINES system. If a developer leaves, there would presumably be a reduced contribution of human capital toward development, support, and training. There may also be a reduced level of symbolic capital as the loss increases concerns that the project may not be completed as promised.

Unlike many OSS projects, the ability of the GPLS as the sponsoring organization to hire a new developer from the library developer community reduces the risk of this stress. The external community has a reasonable degree of awareness in the progress and features of the software because of the open nature of open source projects. As stated by the Deputy State Librarian in a memo shortly after the project was first announced, deploying the software under an open source license eliminates being locked into the specific developers.

"[O]ur open system will be owned by GPLS and we will be free to hire other developers and system administrators to further develop or maintain the system if required ... By fostering a worldwide community around our software, we're actually creating outside expertise in the system, and safeguarding our future."

In other words, the system enables resilience by retaining primary control of the code by employing the developers themselves, having sufficient and surplus resources, and maintaining the social contacts (structural social capital) to be able to access the skills necessary to replace the lost capital contributions. Because of the tenets of the GPL license under which the software is distributed, any other party can certainly access the source code and modify it to fit their own specifications with little or no objection from the current administrators. In fact, this would be encouraged in order to benefit from the additional inputs and efforts of the new developers.

It is also beneficial that there are no reasonable alternatives for the librarians in case of any problems with the development process. State law does not require a library system to deploy the current or its replacement. In fact, several of the larger library systems in the state do not participate in this consortium. As such, it is possible that an individual librarian could conceivably secede from participating in the current consortium and decide not to utilize the new system, but a replacement would be significantly more expensive (especially compared to the "free" deployment of PINES or Evergreen). Were there another inexpensive, fully supported alternative, the individual library systems might have more incentive to consider abandoning the project altogether. Instead, the lack of such an alternative would induce a commitment by both parties to achieve a working solution together. In the worst case, major portions of the software itself would need to be rewritten, but there is little chance of it being dropped altogether based on the current successes.

Conclusion

In this chapter, the concept of open source software ecosystems was defined and examined within a specific case study. An ecosystem consists of a number of units involved in a number of relationships to produce a set of functions. Further, the health of the ecosystem was conceptualized and studied in the same case. Support was found for three dimensions of ecosystem health in terms of the capital flows and accumulations by units in the ecosystem. Organization consists of the various units and relationships. Vigor is represented by identifying the specific capital contributions and functional outputs within the ecosystem. Resilience is examined in the context of three stresses that may occur within the system: inadequate functional fit by the current vendor, loss of developers, and problematic deployment of the new system. The availability of surplus resources, ownership of the code, access to additional sources of human capital, and commitment to make this solution work are seen as contributors to resilience in the current case.

The details of the case study also highlight several aspects that are specific to open source projects that are sponsored by a single organization. Unlike other types of open source projects, the sponsorship both constrains and encourages innovation. The constraints are based on the singular focus on a single user's needs, which is understandable since that user is providing the resources that ultimately lead to its creation. However, the sponsorship also has a number of advantages, including supporting resilience as discussed above.

The mechanisms which lead to the increased vigor, organization, and resilience may prove to be different for other types of open source projects (e.g., community-based or professional open source) as briefly discussed earlier in the context of the PostgreSQL community. Further research would need to apply and further refine the framework to fit these other communities. Additionally, quantitative indicators of each of the dimensions of ecosystem health need to be developed and adapted for open source software ecosystems. Measures of these dimensions have been established for profit-seeking organizations (Iansiti & Levien, 2004), but many of these indicators are based solely on the attainment of financial returns, which does not apply for many OSS communities. As such, the conceptual framework developed above has potential for future research efforts and refinements for both theory development and theory testing.

References

Adler, P. (2001). Market, hierarchy, and trust: The knowledge economy and the future of capitalism. *Organization Science, 12*(2), 215-234.

Becker, G. S. (1993). *Human capital: A theoretical and empirical analysis, with special reference to education* (3rd ed.). Chicago, IL: University of Chicago Press.

Bourdieu, P. (1985). The forms of capital. In J. G. Richardson (Ed.), *Handbook of theory and research for the sociology of education* (pp. 241-258). New York: Greenwood Press.

Chapin, F. S., Matson, P. A., & Mooney, H. (2002). *Principles of terrestrial ecosystem ecology*. New York: Springer-Verlag.

Coleman, J. S. (1988). Social capital in the creation of human capital. *American Journal of Sociology, 94*, S95-S120.

Costanza, R., & Mageau, M. (1999). What is a healthy ecosystem? *Aquatic Ecology, 33*(1), 105-115.

Diamond, J. (2005). *Collapse*. New York: Viking Penguin.

Evans, F. C. (1956). Ecosystems as the basic unit in ecology. *Science, 123*, 1127-1128.

Hannan, M., & Freeman, J. (1977). The population ecology of organizations. *American Journal of Sociology, 82*(5), 929-964.

Hars, A., & Ou, S. S. (2002). Working for free? Motivations for participating in open-source projects. *International Journal of Electronic Commerce, 6*(3), 25-39.

Hawley, A. H. (1986). *Human ecology*. Chicago: University of Chicago Press.

Hertel, G., Niedner, S., & Herrmann, S. (2003). Motivation of software developers in open source projects: An Internet-based survey of contributors to the Linux kernel. *Research Policy, 32*(7), 1159-1177.

Horne, J. F., III. (1997). The coming age of organizational resilience. *Business Forum, 22*(2/3), 24-28.

Iansiti, M., & Levien, R. (2004). *The keystone advantage: What the new dynamics of business ecosystems mean for strategy, innovation, and sustainability*. Boston, MA: Harvard Business School Press.

Krishnamurthy, S. (2005). An analysis of open source business models. In J. Feller, B. Fitzgerald, S. Hissam & K. Lakhani (Eds.), *Making sense of the Bazaar: Perspectives on open source and free software*. Boston: MIT Press.

Lacy, S. (2005, October 7). Open source: Now it's an ecosystem. *Business Week*.

Lin, N. (2001). *Social capital: A theory of social structure and action*. Cambridge, UK: Cambridge University Press.

Lindeman, R. (1942). The trophic-dymanic aspect of ecology. *Ecology, 23*(4), 399-418.

Mageau, M., Costanza, R., & Ulanowicz, R. E. (1995). The development and initial testing of a quantitative assessment of ecosystem health. *Ecosystem Health, 1*(4), 201-213.

McKenzie, R. D. (1934). The field and problems of demography, human geography, and human ecology. In L. L. Bernard (Ed.), *The fields and methods of sociology*. New York: R. Long and R. R. Smith.

Nahapiet, J., & Ghoshal, S. (1998). Social capital, intellectual capital, and the organizational advantage. *Academy of Management Review, 24*(2), 242-266.

O'Mahony, S. (2005). Nonprofit foundations and their role in community-firm software collaboration. In J. Feller, B. Fitzgerald, S. A. Hissam, & K. R. Lakhani (Eds.), *Perspectives on free and open source software* (pp. 393-413). Cambridge, MA: The MIT Press.

Pickett, S., & Cadenasso, M. L. (2002). The ecosystem as a multidimensional concept: Meaning, model, and metaphor. *Ecosystems, 5*, 1-10.

Selye, H. (1974). *Stress without distress*. New York: Lippencott.

Sutcliffe, K., & Vogus, T. J. (2003). Organizing for resilience. In K. Cameron, J. E. Dutton & R. E. Quinn (Eds.), *Positive organizational scholarship* (pp. 94-110). San Francisco: Berrett-Koehler Publishers.

Tansley, A. G. (1935). The use and abuse of vegetational concepts and terms. *Nature, 16*(3), 284-307.

Vecchio, D. (2004). The future of the changing IBM mainframe ecosystem. *Gartner Research COM-22-1194*. Retrieved August 2005.

Watson, R. T., Wynn, D. E., & Boudreau, M.-C. (2005). JBoss: The evolution of professional open source. *MIS Quarterly Executive, 4*(3), 329-341.

Woods, D., & Gauliani, G. (2005). *Open source for the enterprise*. Sebastopol,CA: O'Reilly Media, Inc.

Yin, R. (2002). *Case study research, design and methods* (Vol. 3). Newbury Park: Sage Publishing.

Youndt, M. A., Subramaniam, M., & Snell, S. A. (2004). Intellectual capital profiles: An examination of investments and returns. *Journal of Management Studies, 41*(2), 335-361.

Endnotes

[1] This is not precisely true. Firms such as MySQL allow users the option of paying a propietary license fee in order to avoid releasing derivative products via open source. The practice of offering both open source and proprietary versions of software is known as dual licensing.

2   http://www.postgresql.org

3   See http://blogs.ittoolbox.com/database/soup/archives/innodb-and-the-com-
     promise-of-dual-licensing-6068 for a brief discussion of the effect of the
     GreatBridge era for the PostgreSQL community by one of the core team
     members.

4   This is strictly a hypothetical exercise. I certainly know of no immediate reason
     that any of the developers would be considering leaving at this time.

Chapter XII

The Rise and Fall of an Open Source Project:
A Case Study

Graham Morrison, Linux Format magazine, UK

Abstract

The majority of open source projects fail. This chapter presents one such project as a case study, written from the perspective of the sole developer. It charts the various stages of development, from initial motivation and enthusiasm through the later stages of apathy and decline. It deals with many of the problems encountered by a sole developer, and the various approaches undertaken to maintain development momentum. This chapter provides anecdotal evidence as opposed to statistical analysis, giving an individual's perspective on the development life cycle of an open source project, illustrating real world barriers to development and the typical issues that can stall a project.

Introduction

Where and why does an open source project start? What motivates the developers, and drives them to create applications and tools that are often equal to software many users are accustomed to paying for? Why are there so many that do the same thing? And why are the vast majority of projects left languishing, unloved and un-patched, forgotten and abandoned in some corner of the Internet? This chapter deals with these questions through the eye of a developer who started one such project—a modest digital photo and image manager called Kalbum, which briefly flourished between October 2002 and March 2003. Kalbum itself is suffering a slow and prolonged death, tucked away somewhere on a hard disk in the vast array of storage hosted at Sourceforge.net. SourceForge.net is a testament to the power of open source, as well as its impermanence. It's a Web site that hosts well over 100,000 open source projects, and yet only a tiny fraction receives any kind of attention. For every amazingly successful project like Firefox, there is a Kalbum somewhere, lurking in the shadows.

Background

The dynamics of a project like Firefox and a project like Kalbum are entirely different. For one thing, Firefox has never needed to depend on a sole developer, never had to fight for time with work and family commitments. Firefox inherited much of its codebase from Mozilla—one of the largest, and incumbent community-developed projects in existence. More surprising perhaps is that there is anything common to Kalbum and Firefox other than the open source development model, and, of course, there is. The motivation for Dave Hyatt and Blake Ross to fork development (Hudson, Morrison, & Veitch, 2005) from what they saw as Mozilla's lumbering, over-burdened Web browser is exactly the same as the spark of motivation that ignites the first coding session, the first night of furious programming, that forms the basis for a new project.

This is why Linux has become so successful. While undoubtedly benefiting from the main tenets of open source freedom, as laid down by Richard Stallman in his 1985 *GNU Manifesto* (Stallman, 1985), it's the motivation of its thousands of developers that gets the work done. And there are a thousand different reasons for each developer's motivation, whether it's simply because they're being paid a wage by a company like Red Hat, who need the Linux kernel tailored to their requirements, or an individual who just desires a certain device to be supported. The freedom provided by Stallman's license creates an environment of development unlike any other, but it took the inclusive community that grew around Linus Torvald's kernel to begin to make a difference.

Even today, there is a marked difference between Stallman's religious fervour and Torvalds' more pragmatic approach. Stallman would rather we use free software at all costs, whether or not it might be inferior to a proprietary product. Torvalds, on the other hand, doesn't wish to exclude anybody from the table. A recent illustration of this difference in opinion between the two open source heavyweights is their attitude towards digital rights management (DRM). DRM is used to restrict media content, such as DVD playback and downloadable music, so that it cannot be copied and used freely. The manufacturers behind DRM insist that it's their only defence against a burgeoning problem with piracy. Typically, DRM encrypts data as it travels across insecure protocols, whether the Internet or a cable between a media player and a screen. Only trusted components, hallmarked by the authority that protects the media content, are allowed to decode the media.

Any modern operating system is going to have to support various DRM mechanisms to be able to provide playback of the media content that users will expect. The problem with an open source operating system is that, under the currently proposed terms of version 3 of the GPL, the codes that are used to encrypt media content will need to be included alongside the source code. This is just what Stallman wants, because it will dissuade media companies from using Linux, and distance open source from DRM. He'd rather DRM be referred to as digital restrictions management, an invention that he believes can only curtail people's freedom.

Sony BMG received a great deal of bad press in October 2005 (Electronic Freedom Foundation, 2005), when a certain number of its audio CDs were found to surreptitiously install Sony's own low level control software onto computers running Microsoft Windows. This software disabled a user's ability to make digital copies of the audio—as you can with a regular CD that adheres to the Red Book standard. But how useful is any modern operating system going to be to its users if they can't play the media they want? This is more in line with Torvalds' opinion, which is inclusive for all developers, no matter how they choose to use the technology. If two developers as prominent as Stallman and Torvalds can take opposing sides in such a vital detail, it's no surprise that the motivations for open source development can at best only be described by the developers of any one project.

It's from this angle that we come back to the drive behind this chapter—to find out what motivates a developer and why the project begins in the first place, and what drives its development. Kalbum is still used, it's still downloaded and included with various Linux distributions, but the source code hasn't been touched in over six months. Kalbum is one of many digital photograph management tools available under an open source license. It lets the user manage their collections, as well as add their own comments, change dates and export their collections as a working Web site.

Kalbum isn't a significant project, but it does represent the classic lifecycle of an open source project. From fiery birth, rapid growth, to community development and expanding distribution. Finally, there is developer apathy, slowing development, and boredom. There is a point somewhere in this cycle where the project could have succeeded; an event horizon, where, if there was enough initiative, the project would be pushed on to the next stage—bringing new development, new users. And Kalbum never quite made it to that point.

Motivation

Despite many developers having high ideals and worthy philosophies, many small projects are designed to scratch a developer's itch—whether that's learning a new programming language, toolkit, or operating system. But for the majority of us, it's simply to fill a hole in functionality. Something is needed to perform a certain task that isn't easily available in any other way. This is certainly true of Kalbum. In 2002, digital photography was just becoming affordable for ordinary users. Digital cameras were no longer blighted by low resolutions, poor image quality, and costly storage. It was the cusp of the digital camera revolution, and people were buying new cameras in the thousands.

Linux itself was also in a waxing phase. I'd been using it for a few years, and it was just getting to the stage where I, and many other people, found ourselves using it more and more. This was thanks to the two main desktop environments reaching a certain degree of maturity and functionality. With Linux, the desktop sits on top of the rest of the system, independent of any of the other processes. The windowing system handles the hardware, as well as many of the primitive drawing routines. Because of this independence, it is possible to stop and start the desktop environment autonomously, without restarting the whole system and it's also possible to choose between different implementations of the desktop environment.

Choosing a Development Environment

There are easily more than half a dozen popular desktop environments, but the largest user share is split between two—GNOME and KDE. Typical of many open source projects, GNOME and KDE have both developed from conflicting viewpoints. KDE came first. It was created in 1996 by Matthias Ettrich, a student at the University of Tübingen in Germany. The UNIX desktop at the time was awash with many different userinterfaces and the APIs programmers use to create them. Very few applications looked the same, or shared any of the same functionality we take for granted in a modern desktop. File requesters, help files, configuration utilities, and menus would all be different, depending on how the mouse and keyboard were configured. This disparity contributed to a user experience that fell far short of the equivalent Windows 95 or Mac user experience. Ettrich recognized that this needed to change, and that Linux needed a unified API for developing applications with the same look and feel, and without the developer having to re-invent the wheel every time they wanted to add a certain feature. He outlined his proposal for a unified desktop in a famous posting to a German newsgroup (Ettrich, 1996). The post was titled "New Project: Kool Desktop Environment. Programmers wanted!" In this, he asked what a GUI should be, and listed the common solutions available at the time, and how he thought they had failed. What Ettrich wanted was a complete graphical environment that would allow users to perform the most common tasks, like starting applications, checking e-mail, file management, and configuration, from the same kind of interface. All parts needed to fit together and, most importantly, he wanted this new desktop for the end user—not the system administrator. Linux had reached a threshold where developers like Ettrich were moving to act on their beliefs.

Ettrich had selected a widget library called "Qt" as the platform for his new desktop. Like Torvalds' attitude toward DRM in the Linux kernel, Ettrich didn't base his decision on a political agenda. Instead, he found in Qt a portable, well designed API that he considered ideal. This was a contentious decision because, at the time, Qt was a commercial toolkit released under its own proprietary license. This meant that it wasn't freely distributable, and was in complete contrast to how many people were working. More importantly, it was seen as a serious obstacle to collaborative development. Who would own the resulting code if it wasn't released under a truly free license? However, the desire for a better desktop was stronger than the desire for software freedom, and these problems didn't get in the way of KDE's success. Ettrich's newsgroup post attracted many like-minded developers who wanted to change the face of the UNIX/Linux desktop and, by early 1997, KDE applications were starting to appear.

But the Qt license issues didn't go away, and members of the GNU project were particularly concerned that GNU/Linux might become encumbered by a license that didn't adhere to their strict definition of what constituted "free." Qt was free beer, not free speech—a popular open source expression. Their response was to start their own project, something that would achieve the same end as KDE (there was obviously a desperate need for a unified desktop), but without sacrificing any freedom in return. Two developers, Miguel de Icaza and Federico Mena, picked up the gauntlet. First, this meant developing a toolkit that would rival Qt and, second, building a suitable desktop environment to compete with KDE. GTK+ became the toolkit, and the resulting desktop was called GNOME—the official desktop of the GNU Project. GNOME was, and still is, the yin to KDE's yang. It's built predominantly using the "C" programming language, although C# is becoming increasingly popular, and GNOME applications often take a less-is-more approach to interface design. KDE applications are built predominantly using C++, and they don't shy away from adding one option after another to the user-interface. Of course, competition between the two desktops is often seen as a good thing, but there is a great deal of redundancy. For example, each implements functionally identical libraries for handling vector graphics and desktop searching. The only real difference is their philosophy.

Getting Started

By 2001, I was thinking of starting development on my photo management tool. The first decision was which desktop environment to use. At that time, KDE was more advanced. The biggest obstacle to developing with KDE had recently been removed. KDE had proved to be overwhelmingly popular, to such an extent that it had become by far the biggest project using the Qt toolkit. Qt's parent company, Trolltech, had even employed Ettrich to work on KDE as well as make contributions to Qt full time. After several years of vocal opposition to the license under which Qt was released, Trolltech finally relented and, in September 2000, released Qt under the GPL on UNIX systems (Trolltech, 2000). This is the same license as the Linux kernel, and satisfied many of the dissenters in the community, instantly making Qt/KDE a legitimate development tool for open source developers.

At the time, GNOME was still in its first version revision (1.4), and KDE was well into its second revision (2.2). As a result, I felt KDE was the more mature tool, and offered a greater feature set. This, combined with Qt's license change and a desire to put what I knew about C++ into practice, were the primary influences on my decision to choose KDE as the desktop backend for my photo management application.

Moving to C++ meant that I would finally be able to use some of the object-oriented theory that I'd been reading so much about. All my previous projects had been using C, and, common to many bedroom programmers, I'd not finished any of them. My interest in programming was more theoretical—I liked splitting a problem into smaller functional components, and the challenge of getting them all to work with one another. I understood how this breakdown of functionality could work in an application based on C, but it was the idea of attaching this functionality to distinct objects that really intrigued me. With C++, this functionality can also be inherited by objects derived from a parent object. Qt and KDE use this approach for everything. The main window of an application is a single class of object, and all functionality is passed down to its children. This degree of re-usable code seemed like a big advantage over GNOME, which was firmly entrenched in using classic C functions and libraries. It also meant I would learn something new.

The second version of KDE was a stable and well documented platform, and even featured its own integrated development environment—an early version of KDevelop. Using a template provided with KDevelop meant that it was a 10 minute task to get an initial application framework compiled and running. This really helped to get an overview of how KDE applications were bolted together, as it included a file requester, toolbar, menu, and help system. The template source code was also relatively well documented, and it was easy to see where you needed to start adding your own functionality.

This was one of the most important aspects to the early stages of Kalbum development. I'd never tried creating an application on this scale before, and without a working template, the learning curve would have been too steep. It was actually very difficult to understand how the different re-usable objects within the template source code fitted together. It seemed like they were interacting through the ether, using a form of magic. Without a working template to dissect and experiment with, it's unlikely that Kalbum would have even got off the first step.

The Hurdles

A common mistake for a developer, once they are initiated into the ways and whys of a certain API, is to forget how beginners approach their work. This became the single biggest problem with Kalbum development, and must hold back many other would-be developers from making a contribution. The problem is that, while there is usually suitable documentation for the initial stages of creating a working development environment, as well as getting the first working framework of your application compiled, there is seldom anything to bridge the gap between this template and adding functionality. Obviously, this is difficult, as it's going to be completely different for each application, but with KDE in particular, extra help is needed. The template application you were left with was perfect for programmers who already understood the intricacies of KDE development, but not for those who didn't know the lay of the land, or were perhaps coming to KDE from another desktop environment. What we needed was some simple examples, a guide of how KDE's disparate technologies work together and where to go from the first stage.

As became common practice for me with Kalbum development, it was searching through other applications that provided me with the most answers. This has to be one of the greatest advantages with open source software—finding a tool you know performs the task you're looking for, downloading the source code, and seeing exactly how the developer approached the problem. In the early stages of a project like Kalbum, I was able to piece together the formative application in this way. A perfect example of the difficulty I was having was the toolbar and menu system. These were part of the template, and were working from the earliest stage, but adding further entries to both the toolbar and the menus seemed impossible. Showing users how to add entries to a menu is more important than providing a working template. It was great that the KDE team had gone to the trouble of creating something to help with the early stages of development. But, if that was at the expense of beginners failing to learn vital concepts, such as how the menus and toolbars were populated, it wasn't serving its purpose. The documentation was next to no help with this either. It helped you create the template, and it also documented every feature of the API, as you would expect, but there was nothing to explain basic development considerations when putting together your first application. The answer for the menu and the toolbar was a concept particular to KDE called "actions." These were defined using a specific kind of object, and even after your source code was modified to accommodate them, you would also need to add entries to an XML file somewhere else on the system. In this situation, you might ordinarily expect to find a book that fulfils this requirement. This is a problem with open source.

Despite the best efforts of publishers like O'Reilly and Apress to try and keep ahead of the curve when it comes to emerging open source projects, it can be impossible to keep up with a project as rapidly evolving as either KDE or GNOME. Luckily, there was one such resource, David Sweet's *KDE 2.0 Development* (Sweet, 2000). The book had just been released using an open source license—the Open Publication License, which meant that it was freely available either as HTML or as a formatted PDF. This book was a great help in getting to grips with just how the various KDE components worked together—the most perplexing problem at that point. Sadly, even though the book was fresh from the virtual printing press, I wasn't able to apply the examples to the code generated by KDevelop for two reasons. First, as a perfect example of how two programmers can approach the same problem from two completely different angles, David Sweet's approach to KDE development, at least at the formative stage, was completely different to that of the KDevelop template programmer. For a beginner, it was difficult too see how the two were referring to the same API. The other problem was that even though it had only been a couple of months since the book had been published, KDE had already been through two major revisions—from 2.0 to 2.2, and a considerable chunk of the API had changed. You were never really sure that a problem was either because the implementation had changed or that there was an arcane solution poorly documented, such as with the use of XML for the toolbar and menus. This is a problem that plagues open source development to this day.

The book did work well as a primer of what KDE was capable of, and also for its discussion of certain aspects and capabilities of the API. Without it, I would have had no idea of the signals and slots that Qt used for inter-process communication, another vital technology that's essential for KDE development. This was described in the online beginners guide provided by the volunteers that documented KDevelop, but it took the words of David Sweet for me to be able to understand the concept.

Signals and slots are technologies that makes Qt and KDE so easy, by allowing you to send messages between the user-interface, application logic, and other KDE tools without resorting to buffers and polling. It's an excellent implementation, and so obviously brilliant that it wasn't worth the KDE developer's time to inform the world of just how good it is, and how it's going to make things easy for the KDE developer. This is another common problem with open source software, and later versions of Kalbum in particular took this to heart.

Rapid Development

After the initial teething problems of knowing just where to start, Kalbum went through a stage of rapid growth. When I'd learned how to make additions and modify the template, there was no stopping me. The first thing to be bolted onto the template was the thumbnail image view. This was easy, thanks to KDE's implementation of just such a view—the one used to browse files and images in the desktop's file manager. But KDE itself had also inherited the view from the Qt API, which in fact featured its own version of the icon view I was using for browsing thumbnails. There is hardly any difference between the two, and no clear reason why a separate class of icon view was necessary in KDE—the signals and slots the programmer used to communicate with the icon were identical. The truth is that the KDE development team seemed to err on the side of caution when creating its own implementation of Qt objects. The idea is that they can add their own branding and functionality, and there are many instances where they thought this was necessary. But the icon view was the first example I came across where the KDE team had added very little, and were in fact simply reserving their icon interface for the future. If the worst happened, and KDE had to change the toolkit it was using from Qt to something else, many of the inherited objects, which KDE applications used as an interface to Qt functionality, wouldn't need to be changed. This might sound like a far-fetched example, but in the early stages of KDE development, when the Qt license was causing all kinds of trouble within the community, this is exactly what was happening. Rather than recreate the whole desktop environment and toolkit, as GNOME developers were doing in response to Qt's proprietary license, other developers started work on the Harmony project, a widget-toolkit that would be API-compatible with Qt. Using Harmony instead of Qt could be simplified by only accessing Qt functions from the KDE API, rather than each application using a mixture of Qt and KDE objects. This is why so many objects like the icon view were recreated in KDE, when they added very little to their Qt counterpart. But Harmony development became redundant when Qt was finally released under the GPL.

This wouldn't have been an issue if it was transparent to the developer, but it could create other problems. I was using a tool called QtDesigner to design the user interface, and while it was easy to add and use KDE widgets, it was easier still to stick with Qt's own. This created problems when I needed to switch to the KDE version. I needed to edit the preprocessor "moc" files that create all the user-interface objects by hand before Kalbum would compile correctly. These files basically filled out all the macros used within KDE and Qt objects, and were another arcane addition to the API that was difficult to understand at first; "moc" files were never intended to be edited by hand.

Unfathomable Moments

Developer naiveté had led me to assume that because I could drag and move thumbnails as part of the built in functionality of the icon view, it would be a trivial task to enable Kalbum to re-order images in the main view this way. I was wrong, and this proved to be the most complex part of Kalbum—which is crazy when you consider the simplicity of the idea behind it. I wanted the user to click and drag a thumbnail of the image they wanted into another position, and for the other thumbnails to re-arrange themselves accordingly. The only functionality the icon view actually offered was simple sorting, which was more in line with its original design as the icon view for a file manager. The only solution I could think of for reordering the list of images was to remove the selected images, and insert them either to the left or the right of their original location. The only exception to this rule was if the selected images were either at the beginning or the end of the collection. In these exceptions, the thumbnails would need to be taken from the beginning and added to the end or vice-verse—taken from the end and inserted at the beginning. I was digging myself into a hole at this point. I was cutting and pasting multiple thumbnails from the list of images, and making exceptions for their re-insertion into the album when an image was at the beginning or at the end. For a brief moment I could see the solution, and quickly typed out the code to execute this. It worked, and was an incredibly concise piece of code for changing the order of the image thumbnails within the icon view. The problem was that I only understood the logic behind the code for the few seconds I was typing it out. The moment I'd finished, I was unable to work out exactly what was happening, and I've never understood it since. The code still works, but it could be spreading a virus throughout the Internet for all I understand about how it works. As such, it would be the ideal place for malevolent instructions, and very difficult to find if the primary developer can't even decode his own logic. This has surely got to be one of the biggest threats to open source software—after all, Kalbum and thousands of applications like it are packaged and bundled with dozens of Linux distributions, straight into the hands of unsuspecting users. But the code did the job, and for the most part Kalbum was shaping up nicely.

Preparing for Release

A lot of effort went into the user-interface, creating the text-based file format and even the various methods for sorting the images in the thumbnail view. I'd also added a separate window to let the user describe each photo by giving it a date, as well as adding their own comment and an image title. This would all be saved for reference in the new file format, but there was very little else you could do with the application. There needed to be some way to get the images out to the wider world.

And the answer, of course, was the World Wide Web. This is the perfect platform for people to share their images, a fact that hasn't escaped some of the new breed of Internet sites such as Flickr, which hosts a community of image sharing photographers. I needed a quick solution, something that I could bolt onto Kalbum without too much thought. I found designing Web sites a particularly laborious job, and didn't feel particularly inclined to go to the trouble of creating a new one for the sake of my application. Fortunately, I had a couple of rough Web sites that I had developed for a different purpose, and one in particular fitted the bill. It was a simple design that was based on a tabbed folder, and with a few simple modifications to the HTML, it was turned into a simplified photo album. There was a front tab page for the title, date, and a brief description, a second tab page that almost emulated the thumbnail view of the main application, and finally a third tab for getting a closer look at each image.

Adding the ability to export the HTML from the main application was relatively straightforward. I removed all of the content from the Web page's design, and replaced it with simple tags embedded with HTML comments. Thanks to the excellent text processing functions within Qt, it was an easy task to go step through the skeleton of HTML pages—one each for the title page, thumbnail view, and image detail, and embed the data that the user had added to their collection. The user also had some control over how the HTML folder looked, as I added an option to change the color of each page by simply changing the hue.

The HTML output from Kalbum looked quite good. It was a simple design, low on bandwidth and easy to navigate and understand. Despite the text on the tabs being in English, it should be easy enough for any nationality, and level of experience. I think this is central to the success of Kalbum. Users were more attracted to the quality of the output, rather than the functionality of the application. Using Kalbum was a means to an end. It's easy to see just how diverse the user group of Kalbum became by performing a simple Internet search for "Generated by kalbum," which is text that's embedded into the title page by default. You will get a list of albums that Kalbum users have made available to the rest of the world, and it encompasses many geographical and political boundaries, and covers almost every subject imaginable.

Cutting Corners

The HTML that Kalbum generates was never meant to be a permanent solution. It was a quick fix to generate some meaningful output from an application that until that point had only allowed the user to play with their images. The HTML isn't pretty, and, as with many other aspects of a project like this one, it's surprising just how many stop-gap solutions, made as a quick fix, not only find their way into the finished application, but go on to be increasingly difficult to eradicate.

Lessons Learned

Kalbum was the result of zero design, and in part an early learning sandbox for a keen beginner at C++, including misinterpreted Qt and KDE technology, and re-used code. For example, as a C programmer, I had been used to simply bolting on functionality as required. In the early stages of Kalbum development, this meant that I simply added new methods to a catch-all class—the "kalbum" class. This had grown to include everything from the freshly written HTML exporting code, to handling the configuration files, and even the thumbnail sorting. This was clearly wrong, and went against everything a neatly designed object-oriented project should have been. But considering the circumstances of Kalbum's development, how else could it have happened? I'd spent several years reading about object-oriented design and implementation, and what I really needed was to get stuck into a project of my own. Kalbum was the project, and, short of removing almost all the fundamental functionality and starting again, there was no way of rectifying the situation. The solution would have been to properly design the application before I started development. Had this been necessary, I'm sure Kalbum would never have even started. There was no way, without discovery though experience and implementation, to divine just how the KDE API actually worked. If I had to sit down and work out the relationships on a piece of paper, I would never have been able to get it to work. This seems typical of many open source projects, such as the KDE's development tool KDevelop. There is an initial, rapid stage of development that grows exponentially against any form of design that can be applied to it. A single developer will hack away with a vision, adding all kinds of features to show to the world, without any regard of how it's going to fit into the bigger scheme of things. If other developers contribute, these patches are merged into a chaotic development tree. Only when there are enough features to make a release worthwhile can the developers find the time to patch any of the more obvious bugs, and perform any degree of testing. This is why so many initial releases, at least in a new area, feel so top heavy. They're overloaded with functionality and burdened by their design. If the project is a success, like KDevelop, there comes a point where the development team says that enough is enough. Armed with better knowledge of their users, the development environment, and the needs of their application, it's often best to simply start again. This is exactly what the KDevelop team decided to do at the end of the 2.0 cycle. The 1.0 framework just couldn't support the ambitious ideas for the next version, and a complete rewrite was required. With open source software, perhaps the first version should be accepted as part of a successful project's life cycle. The initial release should be thought of as a prototype, a proof of concept. If it's deemed a success, and the users are happy with the overriding functionality and ideas, the prototype is deconstructed and replaced with a framework that can properly accommodate future growth, as well as other developers who may be drawn to the project.

Of course, there are many exceptions, and this mostly comes down to the skill of the primary developer. Those that are experienced in projects of this nature will not make the same mistakes. A good example is the three-dimensional, open source astronomy application, "Celestia." The project's main protagonist, Chris Laurel, is a very skilled programmer and, more importantly, project leader. The project attracts dozens of technical contributions and, because of the scientific nature of the application, the code is always going to be obtuse. Laurel has published strong guidelines that any contribution should follow, including the formatting of the source code and the rules governing syntax. This means that an application with the complexity of Celestia has never needed a major revision, and is constantly evolving. It's also one of the more successful cross-platform developments, with the Windows version of the software proving more popular than the Linux version. This is thanks to exceptional design, and setting standards at a high level for every contribution. This is the same professional work ethic development you can see in the Linux kernel, but small projects like mine take a very different path. Early versions are prototypes and, if the application doesn't make it to the next level, it's the idea, and not necessarily the implementation, that was at fault. Kalbum falls very much into this latter category, and the next step was to prepare the application for a release, to see whether the idea would sink or swim. There didn't seem to be much point in going to the trouble of creating an application like this if it wasn't going to be made available for others to use. The next problem was how to get the message out there, and getting people to actually use my creation. As with almost every other open source project, advertising comes for free, thanks to the incredible support of Web sites that are dedicated to hunting out projects like mine.

Release

At the time, there was an immensely popular Web site for any KDE related release (http://apps.kde.com). You needed to create an account for your application that would then let the people that downloaded your tool leave any feedback they wanted, like a score and personal comments. I spent some time finalizing a stable version of Kalbum, and also created packages for the Mandrake Linux distribution I was using at the time. Providing packages like these meant that anyone who happened to be using the same version of the distribution (and Mandrake was very popular) would be able to install the application with a single click. Without pre-built packages, a potential user would need to have a suitable development environment, as well as all the associated development libraries to be able to compile and install the application manually. This is one step too many for most people, and is certainly too much to ask of desktop users—even those using Linux. I quietly uploaded all the packages to the Web site late one evening, and went to make a cup of tea. When I came back 30 minutes later, I'd received my first bug report. Several early adopters had downloaded my Mandrake packages and encountered the first problem. It wouldn't install, and instead would simply complain about missing dependencies–a nightmare that plagues many Linux applications. What had happened was that, because I was using a certain brand of graphics cards along with proprietary drivers provided by the manufacturer, certain system libraries had been patched, and, as a result, needed my specific graphics card and driver to be able to work. Those that had the same combination didn't have any difficulties, but everyone else was out of luck. The only solution was to rebuild the packages on a machine that didn't use the same graphics hardware and without the drivers. I was able to do this, and upload the second version of Kalbum a couple of hours later. With that done, I shut both the machines down and left them until the following morning.

The next morning, I connected to the KDE applications Web site and was shocked to find that several hundred people had downloaded Kalbum overnight. Not only that, but my e-mail was taking an eternity to download on my old modem. My inbox was flooded with people having problems with my application. This wasn't something I had anticipated. I was obviously pleased with the number of people who had downloaded it, but setting aside valuable time to deal with the problems was never a consideration. Linux runs on almost every hardware platform available—the combinations are infinite. This causes problems, and it's very difficult to try and work out solutions with the limited setup I was working with. I was able to help many people with the more common problems, but the majority were left to either work things out for themselves or give up.

Feedback

Despite the teething problems, feedback had been overwhelmingly positive, and many e-mails were simply saying "Thank you." Other developers started to send fixes for problems that were particularly annoying them, but most surprisingly, I received several language translations within the first few days. I learned that there were people out there who were continually looking for software to translate to their mother tongue. It didn't particularly matter what the software did, it was just a way for them to make a contribution to the open source movement. This was thanks to the way KDE applications are built, making it possible for translators to simply step through an automatically generated text file, adding translations in their own tongue using a piece of software. Within a week, there were translations in French, Italian, and Hungarian.

Driven by Kalbum's early popularity, I spent the next couple of months spending a lot of time with the source code. Not only was this the fastest period of development, it was also time to rationalize some of the functionality that was already part of the application.

Redesign

Up until this point, design had only been given the minimum amount of consideration. Now that people were actually using the software, there were numerous features presumed to be obvious that just weren't clear at all—including the image detail window and the re-ordering of the thumbnails. The answer was a complete redesign of the user interface, replacing modal windows with free floating, modular windows, and adding other windows for an image preview and album overview. It was at this point that I started to actively read the various KDE development mailing lists. This was because I knew the various image-related KDE applications were all storing their metadata, the comments, and custom dates, for example, using a completely different format. It was obvious that we needed a standard so that additions made to one image using a certain application would be understood by another. But it was very difficult to broker some kind of standard. This was because we would all need to put extra work into our applications when the easy option was for each of us to forge ahead with our own path, with a minimal amount of effort. Needless to say, no standard ever materialized.

Collaboration

It was also around this time that I learned of another project very similar to my own. Even its name was almost identical, "KalbumPhoto." I got in touch with the author, and suggested we pool resources to make a better, single application. Unfortunately, he wasn't interested, saying that there were too many differences in the way we'd both approached the same problem. I understood why he wanted to do things his way, especially if he wasn't sure of his own capabilities (as I wasn't!), but it would have been great to have had some help.

I still received many e-mails requesting help, and the main reason why so many people were having difficulties was because there was absolutely no documentation. I had included a simple text file with the source code of the distribution that explained how I thought it should be used, but this wasn't included in any of the pre-built packages. Kalbum had also been packaged by many of the major Linux distributions and, in the same way that I had made Mandrake packages available, they were making their versions available. Some were even including it as an official part of the distribution. As the author, I knew absolutely nothing about these additions unless one of the packagers happened to require some extra information from me. However, the extra users were certainly making themselves known to me with their questions and problems.

Documentation

But it was the lack of any documentation that caused most of the problems. Despite all my effort to add features and functionality, I couldn't bring myself to spend any time documenting the changes. As a result, many of them were simply passed by, or even worse, misunderstood. This is probably the most common problem with open source projects of a certain size. If your project manages to attract enough inertia—and Kalbum never made it this far—there will be people who want to contribute through writing documentation. It's difficult to find a developer who, given the choice, would rather spend their time documenting their application than adding tangible functionality. I went for the functionality. Even though it is relatively easy to add documentation to a KDE project (using a similar process that the translators used), it never happened. Had I bothered, I'm sure it would have returned the investment in the time I saved replying to people with problems.

This is directly related to many of the difficulties I had in developing Kalbum in the first place. I would often need to add a feature that I knew was possible, but could find no way of implementing. In this respect, the KDE API documentation was failing in just the same way that my own documentation was failing. The perfect example was adding the drag-and-drop functionality. I had originally thought this would be a trivial task, after all, thanks to KDE's widgets, it was already possible to drag the thumbnail around the screen. Had I stayed with dragging thumbnails, this would have been fine, but what I actually wanted was more complicated. When a thumbnail was first selected from the icon view, I needed to store the data associated with the image—the comments and dates—so that these could be injected back into the thumbnail when it was eventually dropped. There are a great number of KDE applications that already featured similar functionality, but it was nowhere to be found in the documentation. The solution was to rummage through the source code of the applications that I knew offered the same functionality and try to decode just what was going on. I couldn't just transplant the code into my application (although this is perfectly reasonably with open source applications that share the same license), I needed to really understand the problem before being able to attempt a solution in my own project. The solution was complex, involving streaming data and defining new MIME types for my image, but at least there was a solution. Open source is something of a double-edged sword in this way. Developers don't like documentation, but then you're free to look through their source code for ideas. Another problem I had was with implementing a suitable "clickable" calendar to help the user select the data. I was able to copy the complete "calendar" class from the KOrganizer application, and virtually drop it into the Kalbum source code untouched.

Final Stages

Over a period of six months, Kalbum had seen four major updates, and been downloaded thousands of times from my Web site alone. Web sites designed by Kalbum were popping up all over the Internet, and it seemed to be going from strength to strength. But each addition was taking longer and longer to implement, and I found I was working less and less on Kalbum. I still enjoyed working on it, but it had come to the point where I knew that any real progress would only be made if I tore back many of the over-burdened classes and objects, and rebuilt the application with the benefit of hindsight.

Conclusion and Future Work

And so Kalbum sits in the archives at SourceForge.net. Over the last few years, there have been moments when I've felt motivated enough to add one thing or another. The SourceForge.net version has several important features, but I've not been able to find the time to create a new release. It's almost as if the challenge was to finish something, to create something that worked, rather than an on-going concern. There are many open source projects where the developer who starts the project is no longer working on the code. Had Kalbum been adopted by other developers, this would be the case now. I still have a real desire to continue working on it, but the reality is that other things now take priority.

With the release of KDE version 4, I would like to be able to start again with Kalbum. KDE 4 is a new challenge. It's had large chunks of its code rewritten, and features plenty of new, modern, and exciting technologies like integrated desktop searching and a global configuration. Converting Kalbum's source code to take best advantage of KDE 4 will mean starting again and trying not to make the same mistakes. For a start, it needs good design, and things have changed a great deal since I started Kalbum. Web-based applications are almost as functional as their desktop counterparts, and far more convenient if you happen to use more than one computer. Many people now keep their image collection in one of the many new breed of community-driven Web sites. A simple desktop tool for sending a Web-based photo album to your family and friends will no longer be enough. Kalbum will need to do both.

Developers need to use the applications they are developing. It's no use adding a feature because it's clever, it needs to be useful, and that means making the user's task easier. There needs to be testing and continual feedback, as well as decent documentation. Kalbum made it easier for users to share their photo collections, and future versions must aim to make that task even easier, or users are going to switch to something else. And there's no lack of competition—there are now many open source applications that perform almost exactly the same function as Kalbum.

The best thing about open source development is that it's incredibly rewarding. Anyone can download the software that enables to you develop an application, and your work can span the globe. Everyone benefits: users and developers alike, and that's infectious.

References

Hudson, P., Morrison, G., & Veitch, N. (2005). Take back the Web: Firefox. *Linux Format, 66*, 50-56.

Electronic Frontier Foundation. (2005, November). *Sony BMG Litigation Info.* Retrieved October 2, 2006, from http://www.eff.org/IP/DRM/Sony-BMG/

Ettrich, M. (1996, October). *New project: Kool desktop environment.* Retrieved October 2, 2006, from http://www.kde.org/announcements/announcement.php

Stallman, R. (1985). *The GNU manifesto* (Free Software Foundation). Boston.

Sweet, D. (2000). *KDE 2.0 Development.* Indianapolis: Sams Publishing.

Trolltech. (2000, September). *Trolltech offers a choice in licensing with the addition of GPL licensing for the upcoming release of Qt.* Retrieved October 2, 2006, from http://www.trolltech.com/company/newsroom/announcements/00000043

About the Contributors

Sulayman K. Sowe is a final year PhD student at the Department of Informatics, Aristotle University of Thessaloniki, Greece. Sowe holds a Higher Teachers Certificate (HTC) from The Gambia College, Brikama Campus. He also received a BEd in science education from the University of Bristol, UK (1991) and an advanced diploma and MSc in computer science from Sichuan University, China (1997). He taught physics, chemistry, and mathematics at various schools in the Gambia (1988-1998). He was a lecturer in information technology at the University of The Gambia (2002). He worked at the Department of State for Education, The Gambia, as the director of information technology and human resource development—IT/HRD (1998), as a system administrator and assistant registrar II for the West African Examinations Council (1998-2002), and as a database manager for the Medical Research Council (2002-2003). His research interests include free/open source software development, knowledge management, information systems evaluation, and social and collaborative networks. He is currently working on several projects related to free/open source software financed by Greece and the European Commission Information Society Technologies (IST) Programmes. He has publications in scientific journals, conferences, and book chapters.

Ioannis G. Stamelos is an assistant professor at the Aristotle University of Thessaloniki, Department of Informatics and Teaching Consultant at the Hellenic Open University. He received a degree in electrical engineering from the Polytechnic School of Thessaloniki (1983) and a PhD degree in computer science from the Aristotle University of Thessaloniki (1988). He teaches compiler design, object-oriented technology, software engineering, software project management, and enterprise information systems at the graduate and postgraduate level. His research interests include empirical software evaluation and management, software education, agile methods, and open source software engineering. He is the author of approximately 70 scientific papers and a member of the IEEE Computer Society.

Ioannis Samoladas is a PhD candidate at the Department of Informatics, Aristotle University of Thessaloniki. He holds a BSc (Magna Cum Laude) from the same department. His main research interest is empirical FLOSS engineering. Additionally, he has been involved in other areas, such as knowledge engineering systems and distance certification. Currently, he is involved in two major European Union (IST) projects regarding FLOSS: SQO-OSS and FLOSSMETRICS. His day job is as a secondary school informatics teacher.

*   *   *

Grahame Cooper has been working in the field of applied IT for the past 20 years. He has carried out research into the application of IT in the construction industry for most of that time, including the development of systems to help integrate and improve the sharing of information across construction projects. His teaching during this time has been primarily concerned with software and systems engineering, with a strong emphasis on object-oriented development techniques, which he has been teaching since 1987. He has taken an active interest in open source software and open publishing over the past eight years, including participation in the EU funded Sci-X project.

Hans de Bruijn is a full professor of organization and management at Delft University of Technology. He is also the scientific director of the Multi-Actor Systems research program. His research concerns networks and network-organizations, with a strong focus on governance and management issues in networks. He is the author of a number of internationally recognized books on these issues. His conceptual approach has been applied to many research domains, like open source, frequency allocation, the design of policy instruments, and mediation (e.g., interconnection disputes).

Francesco Di Cerbo received a Laurea (MSc) in computer science at the University of Genova, Italy, in 2004. He is a PhD student in electrical and computer engineering at the University of Genova, Italy. He is involved in several European and Italian projects related to agile methodologies and open source software. His research areas include software engineering, software metrics, agile methodologies, and open source.

Jesus M. Gonzalez-Barahona teaches and carries out research at the Universidad Rey Juan Carlos, Mostoles (Spain). He became involved in the promotion of libre software in 1991. Since then, he has carried on several activities in this area, including the organization of seminars and courses, and the participation in working groups. His research interests include libre software engineering, in particular quantitative measures and distributed tools for collaboration. In this area he has published several papers and is participating in some international research projects. He is also one of the promoters of the idea of a European master program on libre software.

Wendy Ivins is a lecturer in the School of Computer Science at Cardiff University. She received her PhD from Cardiff University, and works in the areas of soft systems, change management, and software development.

Carola Jungwirth is an assistant professor at the University of Zurich and chair of strategic management and business policy. Her research comprises different aspects of governance in OSS communities, leadership without legitimation, and incentives for investments in human capital. Carola Jungwirth has published various papers in reviewed national and international journals and serves as referee for various national and international journals.

Stefan Koch is an associate professor of information business at the Vienna University of Economics and Business Administration. He received a MBA in management information systems from Vienna University and Vienna Technical University, and a PhD from Vienna University of Economics and Business Administration. Currently he is involved in the undergraduate and graduate teaching programme, especially in software project management and ERP packages. His research interests include cost estimation for software projects, the open source development model, software process improvement, the evaluation of benefits from information systems, and ERP systems.

Benno Luthiger works as software engineer and specialist for open source in the IT Services of the ETH Zurich. He studied physics and cultural anthropology. In his PhD research, Benno Luthiger analyzed the motivations of open source contributors. His research interests are social movements and technical innovations.

Steve McIntosh is a former British Army officer and now an experienced management consultant specializing in business process improvement, change management, and information requirements analysis. In 1997, he joined the academic staff of the School of Computer Science, Cardiff University to set up an innovative MSc program in strategic information systems using the ideas and methods of "systems thinking" to provide students with the theory, methods, and knowledge to achieve business benefits from ICT in their organizations. He continues to combine his academic career with practice-based research in business and government organizations.

Martin Michlmayr has been involved in various free and open source software projects for over 10 years. He used to be the volunteer coordinator for the GNUstep Project and acted as Publicity Director for Linux International. In 2000, Martin joined the Debian Project, and he was later elected Debian Project Leader (DPL) as which he acted for two years. Martin holds master's degrees in philosophy, psychology, and software engineering, and started a PhD at the University of Cambridge in January 2004. His research focuses on quality management in free software projects.

Gabriella Moroiu earned an undergraduate degree in mathematics, physics, and education in Debrecen, Hungary. She taught mathematics and physics in Hungary until 1999, when she emigrated to Canada. Gabriella completed her Bachelor of Computer Science degree at Carleton University in 2005. Her thesis was on the relationship between the structure of open source software and the structure of its developer community. She currently works as an embedded software developer in Ottawa.

Graham Morrison is a journalist and open source software developer writing primarily for *Linux Format* magazine in the UK. He also makes regular contributions to other publications including *PC Plus*, *PC Format* and *PC Answers*, and is an active proponent of free software (FS). After graduating from De Montfort University in 1995, Graham Morrison spent four years as a partner in a computer graphics company before moving on to system installation and support. His first experience with free software was Red Hat 5.2, running on a Commodore Amiga in 1998. His desire to create software using an open platform led him to open source development and writing about a subject he feels passionately about.

Omer F. Rana is a reader in the School of Computer Science at Cardiff University, and the Deputy Director of the Welsh eScience Centre. He formerly also acted as a technical advisor to Grid Technology Partners, a company specializing in Grid technology transfer to industry. He holds a PhD in Computing from Imperial College (University of London) and works in the areas of high performance distributed computing, multi-agent systems, and data mining. Dr. Rana has been involved in the program committees for various conferences and workshops in the area of Grid Computing.

Gregorio Robles received his telecommunication engineering degree from the Universidad Politécnica de Madrid (2001) and has recently defended his PhD thesis at the Universidad Rey Juan Carlos (2006). His research work is centered on the empirical study of libre software development, especially from, but not limited to, a software engineering perspective. He has been, or is currently, involved in several projects related to the study of libre software financed by the European Commission IST programmes, such as FLOSS (2000-1), CALIBRE (2004-2006), or FLOSSMETRICS (2006-2008).

Walt Scacchi is a senior research scientist and research faculty member at the Institute for Software Research, and also the associate director for research at the Computer Game Culture and Technology Laboratory, both at UC Irvine. He received a BA in mathematics and a BS in computer science from California State University, Fullerton, and a PhD in information and computer science at University of California, Irvine in 1981. On joining the faculty at the University of Southern California in 1981, he created and directed the USC System Factory Project until 1991.This was the first software factory research project in a U.S. university. During the 1990s, Dr. Scacchi founded and directed the USC ATRIUM Laboratory, focused on investigating the organizational and technological processes of system development, with emphasis on software engineering and electronic commerce. Dr. Scacchi left USC and returned to UC Irvine in 1999. His research interests include open source software development, knowledge-based systems for modeling and simulating complex engineering and business processes, computer game culture and technology, developing decentralized heterogeneous information systems, software acquisition and electronic commerce/business, and organizational analysis of system development projects. Dr. Scacchi is a member of ACM, IEEE, AAAI, and the Software Process Association (SPA). He is an active researcher with more than 150 research publications. He has directed 45 externally funded research projects. He also has had numerous consulting and visiting scientist positions with firms including AT&T/Lucent Bell Laboratories, Software Engineering Institute at Carnegie-Mellon University, SUN Microsystems, Hewlett-Packard, Andersen Consulting, and dozens of others.

Andrew Schofield is involved with teaching and research activities at the University of Salford. After obtaining a BSc (Hons) in computer science and information systems and an MSc in managing information technology, he began conducting research leading towards a PhD in free and open source software. Andrew's research focus is free and open source communities, specifically investigating knowledge and resource sharing aspects and addressing issues such as organisation, collaboration, motivation, and governance. He has written several papers on the topic including work based on empirical research conducted both within the UK and at an international level.

Marco Scotto is an assistant professor at the Faculty of Computer Science of the Free University of Bolzano-Bozen, Italy. He is a PhD student in electrical and computer engineering at the University of Genova, Italy. He is involved in several European and Italian projects related to agile methodologies and open source software. His research areas include software engineering, software metrics, agile methodologies, and open source.

Alberto Sillitti is an assistant professor on the Faculty of Computer Science of the Free University of Bolzano-Bozen (Italy). He received his PhD in electrical and computer engineering from the University of Genoa (Italy) in 2005. He is involved in several projects founded by the EU and the Italian government related to agile methods and open source software. His research areas include software engineering, component-based development, Web services, mobile technologies, agile methods, and open source.

Giancarlo Succi received a Laurea degree in electrical engineering (Genova, 1988), an MSc in computer science (SUNY Buffalo, 1991), and a PhD degree in computer and electrical engineering (Genova, 1993). He is a tenured professor at the Free University of Bolzano-Bozen, Italy, where he directs the Center for Applied Software Engineering. Before joining the Free University of Bolzano-Bozen, he was a professor at the University of Alberta, Edmonton, an associate professor at the University of Calgary, Alberta, and an assistant professor at the University of Trento, Italy. He was also chairman of a small software company, EuTec. He has been a registered professional engineer in Italy since 1991 and obtained the full registration also for the province of Alberta, Canada, while residing there. His research interests involve multiple areas of software engineering, including (1) open-source development, (2) agile methodologies, (3) experimental software engineering, and (4) software product lines and software reuse. He has written more than 150 papers published in international journals, books, and conferences, and is the editor of four books. He has been a principal investigator for projects amounting more than five million dollars in cash and, overall, he has received more than 10 million dollars in research support from private and public granting bodies. He has chaired and co-chaired several international conferences and workshops, is a member of the editorial board of international journals, and a leader of international research networks. He is a consultant for several private and public organizations worldwide in the area of software system architecture, design, and development; strategy for software organizations; and training of software personnel. Dr. Succi is a Fulbright Scholar and a member of the IEEE Computer Society.

Anas Tawileh is currently doing his PhD in information assurance in the School of Computer Science at Cardiff University; he is also interested in the impact of technology on society and development, particularly in developing countries. Anas established the GNU/Linux Syria User Group in 2002, co-founded the Internet Society—Syria Chapter, and started the Arab Commons Initiative. He is also involved in many Free and Open Source Software projects and advocacy groups. His areas of interest include F/OSS, innovation, ICT4D, information security, collective creation of knowledge, and technology management and policy.

Michel van Eeten is an associate professor in the School of Technology, Policy and Management, Delft University of Technology, The Netherlands. He is also leader of the Critical Infra Program of the Next Generation Infrastructures Foundation (www.nginfra.nl). He has published on large technical systems, ecosystem management, high reliability theory, land use planning, transportation policy, Internet governance and recasting intractable policy issues. His recent work as a practicing policy analyst includes advice to the directorate general of telecommunications and post, the Delfland Water District, the Ministry of Economic Affairs, KPN Mobile, Rabobank, and the Civil Aviation Authority.

Dieter Van Nuffel graduated from the Faculty of Applied Economics of the University of Antwerp, Belgium in 2005. He is currently working for the Department of Management Information Systems at the University of Antwerp. He is preparing a PhD on enterprise systems. Related research topics he is interested in are the selection of enterprise systems within small and medium-sized enterprises, the impact of enterprise systems on the role of the IT department, Web services, and open source software.

Ruben van Wendel de Joode is consultant at the Dutch consultancy firm Twynstra Gudde and is an assistant professor in the School of Technology, Policy and Management, Delft University of Technology, The Netherlands. His research focuses on OSS communities. He has published his work in journals like *IBM Systems Journal, Computer Standards and Interfaces, Knowledge, Technology and Policy*, and *Electronic Markets*. His research on OSS communities has been financed by two grants received from The Netherlands Organization for Scientific Research (NWO). He is also the lead author of the book *Protecting the Virtual Commons: Self-organizing Open Source and Free Software Communities and Innovative Intellectual Property Regimes* (2003).

Kris Ven graduated from the Faculty of Applied Economics of the University of Antwerp, Belgium, in 2002. He is currently working at the Department of Management Information Systems of the University of Antwerp. He is preparing a PhD on the organizational adoption of open source server software. Related research interests include the link between innovation in organizations and open source software, and the adoption of open source software by public administrations. He has authored and presented several papers at international conferences on open source software.

Jan Verelst received his PhD in management information systems from the Faculty of Applied Economics of the University of Antwerp, Belgium, in 1999. He is working at the Department of Management Information Systems of the University of Antwerp, where he teaches courses on analysis and design of information systems. He is also an executive professor at the Management School of the University of Antwerp. His research interests include conceptual modeling of information systems, evolvability, and maintainability of information systems, empirical software engineering, and open source software.

Tullio Vernazza received a graduate degree in electronic engineering from the University of Genova in 1971. In 1971 he joined Telettra, in Milano, where he worked as researcher for three years. From 1974 to 1983 he held various courses on computer science fundamentals as contract professor at the University of Genova. From 1983 he has been an associate professor on the Faculty of Engineering of the University of Genova, actual professor of computer architectures and software engineering in the curriculum in "Ingegneria Informatica," member of the steering Committee of the Doctor Program in "Scienze e Tecnologie dell'Informazione e della Comunicazione" since his setting-up (1983). Expert in evaluation and review committees for Progetto Finalizzato Informatica 1 and Progetto Finalizzato Trasporti 2, Professor Vernazza has been the leader of several research groups in projects financed by European Community, MURST, CNR, Società Autostrade, public bodies, and private companies. He has been a consultant for such public bodies as the Italian Ministry for Foreign Affairs, Genoa Public Administration, Genoa AMIU, Liguria Scientific, and Technological Park, and Regione Liguria. In the first years of his research activity, he carried on research on computer architectures, design and implementation of semiconductor components, and design of high performance computers. Later, he directed his studies towards motorway traffic problems and the design of systems for traffic monitoring and control. In collaboration with Professors Morasso and Zaccaria, he created the first Italian hospital robot (this achievement was rewarded with the title of Commendatore della Repubblica), looking after the implementation of the electronic and hardware components. His main research interests are robotics and software production processes and methodologies.

Michael Weiss (PhD, University of Mannheim, 1993) is an associate professor of computer science at Carleton University. Before joining Carleton in 2000, he worked four five years in the telecommunications industry (Mitel Corporation) on agent-based service creation environments. His research interests include software architecture and patterns, service-oriented architectures, and open source.

Donald Wynn, Jr. is a PhD student in the MIS department of the Terry College of Business at the University of Georgia. He also holds an MBA and a BS in electrical engineering. Prior to his doctoral studies, he spent 15 years in telecommunications engineering, network management, and information systems management. His current research interests include open source software ecosystems, technology innovation, and information systems security. He has published in *MISQ Executive, The Journal of the Academy of Marketing Science, The Journal of International Management, Communications of the AIS,* as well as several IS conference proceedings.

Index